Production Management of Chemical Industries

Kiran R. Golwalkar

Production Management of Chemical Industries

Springer

Kiran R. Golwalkar
Nagpur, Maharashtra
India

ISBN 978-3-319-28251-0 ISBN 978-3-319-28253-4 (eBook)
DOI 10.1007/978-3-319-28253-4

Library of Congress Control Number: 2015958843

Printed on acid-free paper

This Springer imprint is published by SpringerNature
The registered company is Springer International Publishing AG Switzerland

To the workers, design and shop floor engineers, and progressive managements who give priority to safety of all persons associated in any manner with their industries and products to control the environmental pollution, to maximise the efficiency, and to manufacture the products with the best possible quality.

Preface

Chemical industries are very important for our daily lives since we need various products for personal hygiene and health care, fuels for cooking and vehicles, paints and dyes, petrochemicals, synthetic fibres, water treatment chemicals, fertilisers, insecticides and pesticides for agriculture, etc. Chemical industries also make a large contribution to national economies. However, many of these industries produce or handle dangerous and highly polluting chemicals which must be very carefully controlled. Hence, the governments of all countries insist on compliance with the strict rules and regulations which are enacted to ensure the safety of personnel and the control of environmental pollution.

It is necessary that management, and all those shop floor engineers working in chemical industries (who are looking after the process and maintenance) should operate their plants in a safe, pollution free manner along with an all-around efficiency in order to keep the pollution and the production costs in control while meeting the required product quality and delivery schedules for their products.

Some of their main activities are given as follows:

- Market survey to estimate demand for the product mix
- Setting up new production units, expand/diversify running units, or revive idle units
- Ensuring safe operations and maintenance by Hazid and Hazop studies
- Compliance with statutory rules and regulations
- Procuring materials and machinery
- Environmental pollution control by minimising effluents and treating them properly
- Production of chemicals as per specifications of quality control checks
- Dispatch as per the delivery schedules agreed with clients

The author has attempted to present information generally needed by these engineers (and managers) about technical matters in a bullet format (rather than lengthy explanations) for an analysis of their plants and for proceeding further. It is seen that the main points are easily noted when checklists are given in this manner.

It is hoped this format will enable them to quickly prepare reports for plant expansion, diversification, and purchase of new equipment for presenting to the higher management for approval and sanctions.

Suitable explanatory notes instead of lengthy descriptions are added only at appropriate sections (in the relevant chapters of the book).

There are many excellent books available for researchers. However, I have placed more emphasis on safety, environmental pollution control, and energy efficiency while writing the book and hence request the persons carrying out research to always consider these issues while developing new processes and products.

The book is thus meant for use by production managements, shop floor engineers, and for students who wish to join the chemical industries.

Kiran R. Golwalkar

Acknowledgements

I am grateful to my teachers at the technical institutes who taught the basic engineering principles, my senior colleagues for the training given to me at the beginning of my job in a chemical industry, the shop floor working personnel who were very cooperative and gave many practical suggestions, and to the design and shop floor engineers and progressive managements with whom I had opportunities to work.

I wish to especially thank respected Mr. Brian Halm, Project Coordinator and Mr. Michael Luby, Senior Publishing Editor, Engineering at Springer Science+Business Media, New York, USA for the encouragement and help in completing this work.

I am very thankful to Ms. Nicole Lowary, Assistant Editor, Applied Sciences, Springer Science+Business Media for the support and the Experts who have spent their valuable time, have taken the pains to review the book and made very encouraging positive comments.

I also specially wish to thank Ms. Nishanthi Venkatesan, Project Manager, Scientific Publishing Services (P) Ltd. for the great efforts made for preparation of the page proofs and kind support provided.

Sources of My Information

My own experience of working in various chemical industries and consultancy organisations for the erection, commissioning, operation, modernisation, and diversification in India, Kenya, Thailand, and Indonesia, and visits to industrial exhibitions in India as well as discussions with various vendors of equipment during my professional career.

Contents

Chapter 1
Introduction

Chemical industries play a very important role in our life by contributing various products using every day such as healthcare products, detergents, cooking gas, fuel for vehicles, pharmaceuticals, paints and dyes, plastics, petrochemicals, synthetic fibres, water treatment chemicals, fertilisers for crops, insecticides, and pesticides—the list is very large—almost endless. The chemical industries also make a very useful contribution to the national economies.

1.1 Typical Features of Chemical Industries

- They have large thermal and electrical energy changes (consumption/generation).
- Their operations are mostly automatic. Very careful continuous monitoring is necessary in most of them—(but less manpower is required).
- They do not produce much noise generally (except where crushing and grinding are done).
- Effluents are generated in considerable quantities (toxic, harmful substances), and they can pollute the environment since their disposal is not easy.
- There could be fire hazards or long-term health hazards.

There are a large number of chemical industries in operation and new ones are being added all over the world. However, the general perception is that though very useful, chemical industries pollute the environment and could be very dangerous some times.

Governments of all countries are fully aware of this and have made stringent rules and regulations to monitor their operations to ensure safety of personnel and control of environmental pollution.

K.R. Golwalkar, *Production Management of Chemical Industries*,
DOI 10.1007/978-3-319-28253-4_1

1.2 Modern Managements

Modern responsible and successful managements make efforts to operate the chemical industries without any accidents (completely safe for plant personnel as well as outsiders), without causing any environmental pollution (minimising generation of waste and proper treatment of all the effluents) in a highly energy efficient manner and minimising consumptions of raw material while delivering the products as per specified quality and quantity required by the clients.

The revenues generated are used very judiciously for necessary maintenance of safety devices, effluent treatment, process units and machinery; procuring supplies of raw materials and other inputs; paying taxes and duties to the government, welfare of all working personnel, carrying out innovation for improvement and research for new technologies; and return of capital with interest as per terms of agreement with all creditors and stakeholders.

Reasonable amounts are set aside for replacement of old (corroded) process units and worn out machinery, and as a contingency fund for any unforeseen situation. This is very important for a chemical industry because the rate of corrosion is much more than general fabrication shops or engineering industries. The distribution of profits to the promoters/owners (and the equity shareholders) is generally done after above allocations.

1.2.1 Important Management Activities for the Chemical Industry

Modern managements with a progressive and practical outlook apply their knowledge, experience, and resources for executing their main functions of planning, organising, staffing, coordinating, directing, and controlling the activities essential for the successful establishment, operation, expansion, modernisation, and revival of idle units.

They are able to perform these important managerial functions by employing professionals with knowledge and experience in their respective fields.

There can be a certain overlap among these functions and the activities during execution. Though the activities are allotted to different departments, no sharp divisions are generally made. All departments and sections are supposed to work as one well-oiled machine with clear communication and cooperation with each other; though some demarcation could be made for convenience of working by using the individual expertise for meeting the aims of the organisation.

Broadly these activities are arranged to suit the nature of business (manufacturing or trading), geographic area of operations, whether running on a low-volume, high-unit-price basis or high-volume, low-unit-price basis; dependence on local or imported technology, or materials; producing items for mass consumption; healthcare products, or industrial use; producing goods which need safe and careful handling, and so on.

The main activities typically required for setting up a chemical plant are as follows:

- Market studies and finalising the product mix,
- Basic and detailed engineering (*please see below*),
- Hazid and Hazop studies,
- Incorporating safety devices and revising P and ID, operating procedures, etc.,
- Obtaining statutory permissions and necessary licences,
- Creating site infrastructure and carrying out site activities,
- Procuring plant and machinery,
- Erection and commissioning of the process plant,
- Stabilising safe and efficient operation and maintenance of the plant,
- Minimising generation of effluents and ensuring their proper treatment in-house or disposal through common disposal facilities run by statutorily authorised parties,
- Production of chemicals as per specifications given by customers,
- Quality control checks before dispatch,
- To meeting the delivery schedules agreed with clients.

Production management has to look into all these and ensure that the plant runs safely, without causing any environmental pollution while complying with all statutory rules and regulations, and generates sufficient revenues to get a satisfactory rate of return. These essential activities are briefly described below and elaborated elsewhere in this book.

1.2.2 Estimation of Demand for the Products

An initial estimation of demand for various products is carried out to shortlist them. This is necessary for planning a new industry or for diversifying to additional products from the existing running plant. It can be entrusted to a professional agency if it cannot be done by the organisation itself. It should be a reasonably reliable estimate of demand for the products and their specifications generally required by clients. The products are shortlisted accordingly and the production planning is based on these estimates.

1.3 Basic Engineering Documents

It shall be prepared for deciding further in the matter (*as it gives more information*):

- Production capacities to be planned for the product mix (*at present and in future*);
- Process flow sheet (*for selecting process and technology*);

- Issues related to safety and environmental pollution control;
- Expected running days per year and stoppage for annual overhaul;
- Rated capacity of the plant for various products (continuous operation) or number of batches per year if it will be a batch type operation;
- The expected total investment required;
- Layout for the product mix (*at present and in future*);
- Infrastructure required (land, approach roads, storage);
- Raw material and utilities consumptions (water, power, and steam);
- Engineers, technicians, and workforce required;
- Maintenance requirements and other expenses;
- The expected **return on investment**;
- Time required to implement the project;
- Shortlisted locations based on availability of raw materials, power, water, geographic and climatic conditions, proximity to clients, existence of approach roads, etc.;
- It is important to look into the suitability of climate for safe storage, handling of various raw materials, the products, and safe operation of the process plant. The decision to set up the process industry or for adding more products to existing industry can be taken on the basis of above information.

1.3.1 Rated Capacity

This shall be based on annual production and not on daily production. It should take into consideration the plant stoppages due to annual overhaul, and occasional breakdowns/process problems.

Example
A continuously running plant can produce 100 MT per day of sulphuric acid.

It needs about 20 days of complete stoppage for annual overhaul and maintenance. It is also estimated that the plant would be stopped for a total of about 15 days due to various reasons during the year such as sulphur spray gun choking, air blower maintenance, repair to control valves in acid lines, etc.

This leaves $365 - 20 - 15 = 330$ days of actual running period.

The annual production would be $330 \times 100 = 33{,}000$ MTs and not 36,500 MTs.

The sales volume would be 33,000 MTs, and revenues will be obtained from these only.

However, certain expenses on security, stores management, plant lighting, salaries of supervisory staff, maintenance crew, water supply, etc., will be incurred even when the plant is not producing. Money is spent on all these matters throughout the year for 365 days.

The plant shall be rated as 33,000 MT/annum therefore, and cost of production per MT shall be calculated by dividing total annual costs by 33,000 and **not by 36,500** MT as this can give an incorrect figure.

In case of plants operating with complicated technologies or handling very corrosive products, the running days per year shall be considered even lower—say, 300/310 only.

The profit per MT and return on investment shall be estimated on these figures. It is obvious that all efforts shall be made to minimise the annual shutdown period and other stoppages during the year.

However, the plant shall not be operated at overload condition *beyond safe limits* during the running days to achieve higher production for getting more profits.

1.4 Detailed Market Survey

This is to be done now to obtain reliable information through an experienced marketing organisation. This is necessary before finally deciding to proceed further on the basis of the basic engineering document.

It should cover all the aspects of the demand for the product—the product quality generally specified by clients (minimum purity required, tolerance limits for impurities), the types of containers preferred (carbouys/tankers); quantities of products required and the delivery schedules convenient to them; the present producers of the products, the demand–supply gap in the market, whether demand for the product is seasonal or all through the year, in domestic market or overseas.

1.5 Finalising the Product Mix

- The initial planning for the production unit may be for only one product or more products with a certain capacity for each of them.
- The future product mix (with more products) and the corresponding production capacities can also be planned in the beginning itself while setting up the industry.
- Expansion of production capacities in the future may also be planned in the beginning.
- This will enable design of process scheme and the list of equipments *to be installed at present.* Planning for the site infrastructure for the present and for the future shall also be done accordingly.
- Some of the process units and machinery from the plant can be used **as such** in future also, while some additional equipments may be required for the product mix and production capacities planned for the future. Some existing units may need only upgradation.

A plant for manufacture of sulphuric acid can produce 98.3 % technical grade acid only or can also produce 25 % oleum by addition of an absorption tower with accessories for circulation system. An oleum boiler can then be added to produce

SO_3 vapours for subsequent production of 65 % oleum or liquid SO_3 with some more equipments. Expansion of production capacity can be planned in the beginning by installing sufficiently big sulphur burning furnace, waste heat recovery boiler, converter, air blower, etc. *These can be used as such in future.*

1.5.1 Future Product Mix and Product Life Cycle: A Word of Caution

Very careful assessment of the demand for the product shall be made before planning to manufacture higher quantities by the plant and adding more products in the future. The product quality shall be continuously improved upon, and its usefulness for more applications shall be found out. Feedback shall be obtained from clients by regular meetings to get a hint whether the product may become obsolete due to better/cheaper products likely to enter the market. However, there shall be no compromise on safety, environmental pollution control, product quality, and efficiency of operations during expansion of capacity or while adding new products.

1.5.2 Overloading of Equipments

Equipments such as pressure vessels, boilers, ETP, electrical systems, material handling systems (hoists, EOT, belt and bucket conveyors), storage tanks, cooling towers, condensers for volatile liquids, refrigeration systems, water supply and treatment, pipe lines already operating at high pressure, furnaces and heat exchangers operating at high temperatures, catalysts (due to high input load), agitated vessels (due to high speeds) can get overloaded due to operation at higher production rates.

There can also be chances of maximum electrical demand being exceeded and overheating of transformers, electrical motors, cables, starters, etc., when ambient temperatures are more.

The plant operations shall not become unsafe or environmentally polluting under any circumstances—whether due to more products or due to expansion of plant capacity.

1.6 Selecting the Process and Technology

(*More details are in Annexure A—General Tech. Matters*).

- The one which is most efficient and safe to operate shall be chosen. *It will be advantageous if the climate at desirable location suits the process and technology.*

- Consumption of raw materials and utilities shall be the lowest per unit output.
- Requirement of water shall be minimum (water cooling may not be used if good quality water is not sufficiently available).
- It shall be possible to run the plant efficiently and safely at different production rates.
- Quality of products shall meet the specifications (as per market survey for domestic and international markets).
- Sufficient warning alarms, safety devices, and interconnections must be incorporated.
- The plant should not cause any environmental pollution.
- The process units shall be designed and fabricated as per codes and standards to be statutorily complied with in the country *where the project is being set up.*

The basic engineering document may now be accepted by the management after due deliberations and further work can continue either in-house or by the consultants appointed.

1.7 Detailed Engineering Document

The management decision to move forward and set up the project (new unit/revival of old one) is indicated on acceptance of the basic engineering document. the **detailed engineering document** should give more details of the following:

- Process for manufacture,
- Specifications of raw materials and utilities,
- Guaranteed expected consumptions of raw materials and utilities.

The raw material and energy consumption figures by vendors/technology suppliers to be considered with some additional margin to estimate the cost of production and the profitability before setting up the plant.

- Material and energy balance;
- Plant layout;
- Locations of equipments on various floors;
- Specifications and fabrication details of all the equipments required;
- Effluent treatment and air pollution control facilities;
- Piping and instrumentation diagram (P and ID);
- Sectional elevations;
- Instrumentation details;
- Thermal insulations;
- Civil foundations;
- Electrical diagrams;
- Project implementation schedules;
- Erection and commissioning procedures;
- Operating and maintenance manuals, dos and don’ts;

- Safety precautions;
- Procedures for handling all materials in the plant (raw materials, intermediates, additives, finished products)—unloading, packing; and dispatching;
- Process control and quality control facilities required;
- Manpower requirements and any other information required by the management for the successful implementation of the and running of the plant.

All these are to be studied carefully. Any safety hazards, possibilities of mishaps due to ***unintended*** excess/less/no flow of reactants, high/low temperatures, or pressures in process units, etc., shall be identified and designs rectified accordingly. Additional safety devices may be added.

Readers are requested to kindly refer to chapter on **Safety Management**.

1.8 Procurement of Plant Equipments and Machineries

The [1]process flow diagram, process description, and **P and ID** shall be frozen. The plant layout, list of major process units, and machineries for manufacture and their specifications shall be finalised. This will enable the *procurement activities* to begin for long delivery items. Specifications/designs of minor process unit, auxiliary units for material handling, utilities, storage tanks, packing, weighing and dispatch arrangements can be worked out in the meanwhile and enquiries floated for them also. Short listing of vendors, placing orders, inspection, etc., shall follow till the items are procured and brought to site.

1.9 Creating Infrastructure Facilities

- Establishing site office, logistics;
- Land acquisition;
- Security and living quarters for site engineers, work force;
- Stores (covered shed and open yard) and weighbridges;
- Water and power for construction and fabrication;
- Site fabrication and erection jobs.

[1]*Process flow diagram: Choose such designs and processes where flexibility to produce different products at different rates of production are possible by simple changes in process units or by adding some balancing equipments.*

1.10 Financial Arrangements (Cash Flow Planning)

- Procurement of equipments and site fabrication;
- Fees for statutory licences, and compliance with statutory instructions for any additional items or constructions in the plant;
- Expenses for site office and jobs;
- Expenses for erection, mechanical trial runs;
- Expenses for arrangement for all inputs such as raw materials, spares, fuels, and power till the stabilisation of plant operations till performance guarantee tests are carried out (plant commissioning);
- Power for steady running of the plant for regular production of saleable products;
- Marketing of products at competitive rates;
- Repayments of instalments of loan to financial institutions;
- Managing the expenses till revenue generation starts.

1.11 Organisation of Structure for Industrial Administration

All personnel working in the organisation shall be qualified and experienced to suit the jobs and responsibilities which will be allotted to them. They should have higher-level (university) education from appropriate technical and commercial institutes.

Managing and Controlling of Activities

Directions are to be given to all employees to meet the goals of the company for increased market share, establishing a good reputation for quality, safety, and pollution control; and credibility for meeting delivery schedules, developing new products regularly.

Create an atmosphere of mutual trust through ethical work practices and integrity; and recognition of good work. The policies for work distribution for in-house work, outsourcing certain tasks, and deployment of work force for different work stations (using the personnel by rotation on different process units and machinery to minimise dependence on few key personnel) can then be more easily accepted and implemented.

Different grades of senior executives, middle-level management, shop floor engineers, technicians, and skilled and unskilled workmen can be designed for technical, commercial, and administrative jobs. The relationship for chain of command (superior- and junior-level posts) shall be well defined. Instructions and orders by seniors shall flow down the chain of command only; and reports of compliance, requests for guidance and funds, requisitions for materials and equipments, obtaining approval schemes shall go to superiors through proper channels only.

However, they shall also communicate and coordinate with some other personnel in the organisation who will neither be superior or junior to them. These could be from the R&D, logistics, quality control, stores, and purchase departments. This can be defined as staff function for the smooth working of the organisation.

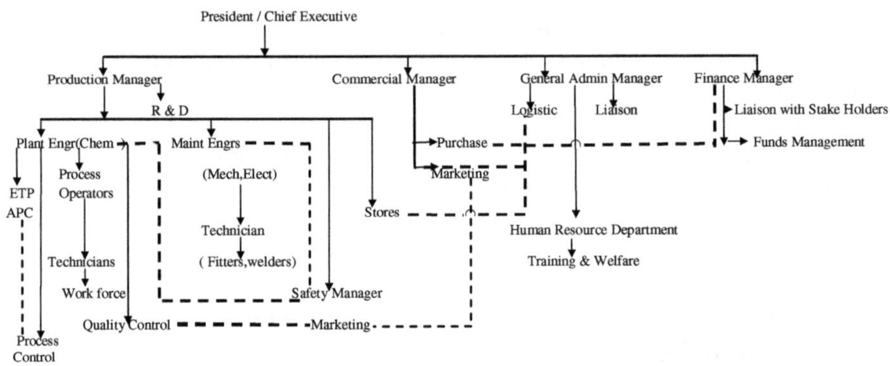

Organisation structure for industrial administration. *Note ETP* Effluent treatment plant, *APC* Air pollution control

1.11.1 Manpower Planning for Various Departments

Plant Operation and Maintenance

- University degree holders for senior positions from appropriate disciplines (chemical, mechanical, electrical, instrumentation, civil engineering) and with sufficient practical experience;
- High pressure licensed boiler attendants for boilers, economisers, superheaters;
- Steam turbine operators—licensed and experienced;
- Licensed electricians (for high tension work), welders (for high pressure vessels), machine and pipe fitters; fabricators, riggers, skilled helpers and unskilled labour;
- Instrumentation technicians;
- Skilled work force trained in technical institutes and who have experience of working in similar industries in plants of comparable capacities and to handle dangerous materials, effluent treatment plants.

Process Control Laboratory, Research and Development

- University degree holders for senior positions from appropriate disciplines (chemical, mechanical, electrical, instrumentation, and environmental engineering) and with sufficient practical experience in material science, metallurgy, applications of organic and inorganic chemistry, knowledge of fabrication and operation of industrial equipments, cost estimation, etc.

Stores Management
University degree holders for senior positions from appropriate disciplines (chemical, mechanical, electrical, instrumentation, and civil engineering) familiarity with process operations, working knowledge of maintenance, and with sufficient practical experience of materials management, and procurement of machinery and their spares.

Safety Management
This is a very important department and only highly experienced engineers having university degree-level qualifications shall be engaged here.

Commercial Management
Sales/marketing, advertising, insurance, purchase of raw materials and spares and **financial matters** taxation, budgeting, salaries and all other payments to employees, payments to contractors, vendors, allocation of resources, stakeholders and their interests, equity, preference shares and debentures, repayments of loan instalments to financial institutes and investors.

General Administration
University degree holders for senior positions with specialisation in human relations, training programmes, welfare activities for personnel, logistics (transport fleet), upkeep of company properties, legal matters, and liaison activities with statutory inspection and licensing agencies.

1.11.2 Statutory Rules and Regulations

All countries have enacted laws for the safety of personnel and property, control of environmental pollution, welfare of work force, financial matters, and taxation which shall be complied with. Technical inspections of process plant and all equipments covered by the rules are also done regularly by statutory authorities. Their instructions are to be complied with.

Prohibited areas:
Establishing the unit in a particular area may not be permitted at all due to presence of schools, hospitals, water bodies such as lakes, rivers; heritage sites or historical structures, and defence or military establishments etc.

1.11.3 Organisation Resource Planning and Allocation

This shall be done for achieving the goal of developing the organisation as an efficient, reliable, manufacturer which delivers quality products as per specifications and delivery schedules committed, while operating safely without causing environmental pollution.

1.11.3.1 Resource Allocation

Maximise utilisation of existing facilities and infrastructure; sometimes by producing on contract for other parties while outsourcing less important items (such as stabilisers, anti-caking agents).

Deputing own manpower to other parties on consultancy basis (for technical assistance or for running their units) can generate more revenues for the organisation, can create more goodwill, and the personnel can get additional *useful* experience.

1.11.4 Running on Contract

Management may consider following options:

- BOT: Build, operate, and transfer. The plant is built and operated by external agency till performance guarantee terms are met and then transferred immediately to the purchaser.
- BOMT: Build, operate, maintain, and transfer to purchaser. The plant is built, operated, and maintained by external agency. At the end of the contracted period, it is to be handed over to purchaser.
- BOOM: Build, own, operate, maintain for some years as per agreement between purchaser and external agency. The profits during these years could be shared or taken by external agency. It is then to be transferred to the original party.

1.11.5 Logistics

The general administration should look after this very important support activity for transport of incoming materials and spares, and dispatch of products. There shall be close coordination with purchase, stores, production, maintenance, and marketing departments.

A list of all inputs to be procured and all dispatches to be made in the next few days/week shall be prepared in consultation with production and marketing.

1.11.6 Communications

Orders and instructions shall be very clear from seniors to juniors, while reports and request from juniors to seniors shall also be clearly understood. Any communications with outside parties shall not be confusing.

1.12 Technical Matters

1.12.1 Procurement of Raw Materials and Other Items

The specifications of all raw materials, *additives*, *preservatives*, spares, and other inputs shall be correctly defined to ensure final product quality, to ensure proper functioning of process units (to minimise scaling on heat transfer surfaces, to minimise system choking due to ash content), to have a long equipment life, to minimise environmental pollution.

Minimum purity required should be clearly written. It is also necessary to mention the maximum limits of impurities, insolubles, contaminants, moisture (i.e. undesirable things) while floating enquiry.

1.13 Safety Management

Safety should be **everybody's concern** during erection, commissioning, trial runs, stabilisation of plant operations, normal running, maintenance, replacement of equipments, expansion of production capacity, diversification to more products, and dispatch of products.

Hazid and Hazop studies shall be carried out to improve the plant design and incorporate additional safety devices so as to minimise chances of mishaps to ensure the safety of personnel and plant equipments. It will be useful if these studies are carried out at every six months interval since some changes (process conditions or operating procedures) may get introduced *inadvertently* by plant personnel over the last few months.

1.14 Process Control Laboratory

These are very essential for assistance during commissioning, analysis of raw materials and intermediate process streams, finished products, effluent streams before and after treatment, and to monitor proper working of the reactors.

1.14.1 Quality Control

The quality of products should be checked before dispatch. It must meet specifications given by clients so that the products are not rejected. Plant design, raw material purchase, process control, instrumentation, and maintenance all contribute towards better quality.

1.15 Utilities

Utilities are required for operation of the main process reactors and related units for pretreatment of raw materials and processing of the products, for safe and efficient operation of plant and treatment of effluents. Main utilities are electrical power, treated water, steam, dry air, heating fuels (mediums), nitrogen flushing for the chemical plant.

1.15.1 Energy Efficiency, Cogeneration, and Various Methods of Heating

All inputs of energy as fuels, electrical power, steam, and the evolution of heat during certain reactions shall be monitored. Equipments shall be installed for heat recovery (for internal generation of power and steam). Various methods for process heating, for reducing energy consumption, and for increasing energy recovery are to be evaluated and put in practice to increase energy efficiency.

1.16 Technical and Other Audits

All activities range from initial design, inspection, and purchase of process plant, and their operation and maintenance; and safe operating practices shall be audited for enforcing proper working. External auditors can be employed for an independent assessment. Regular internal audits for performance, safety, and energy efficiency shall also be carried out.

1.17 Pollution Control and Disposal of Wastes

This should also be everybody's concern ETP. Precautions shall be taken for plant design, operation and maintenance to minimise generation of wastes, recycle of effluents (*if possible*), and treatment of effluents. Disposal of wastes generated during plant operations can be done through suitable paths for disposal of incineration, secured land fill, etc.

1.18 Water Budgeting

Water is a very important resource for many chemical industries. The requirement in the plant shall be minimised by recycling treated water from effluent treatment plant (ETP) as make up to scrubbing system or process units; as make up to cooling towers, for washing reaction units between batches, for floor washing, etc., as much as possible. The water treatment plants and storages of treated water shall be accordingly designed.

1.19 Electrical Installations

Electrical power is one of the most essential inputs for a chemical industry. Requirement of power for plant construction and operation, installations for drawing power from external grid, using it safely and efficiently in the industry, and cogeneration facilities are important.

1.20 Maintenance Management

Condition monitoring, preventive maintenance, and breakdown maintenance shall be well planned, organised, and safely executed in consultation with production team. Some essential tools and facilities shall be available in-house.

1.20.1 Some Quick-Fix Solutions

Some methods for keeping the plant running till proper repairs are done—there are only first aids and *not a permanent solution to the problem.*

1.21 Management of Stores

It is essential to keep all important inputs available for safe, pollution free, and smooth running of the plant; and to enable dispatch of products as per specifications. Certain items can be made in-house or bought from outside.

1.22 Fire Fighting Systems

These are necessary for safety of all process units and installations, safe storage of raw materials and finished products, engineering stores, and electrical equipments. Proper design and provision of all necessary fireighting equipments are mandatory.

1.23 Innovation, and Research and Development

These activities are necessary for more safety in operation and maintenance, better quality of the products, getting leadership in market, for increasing profits, more overall efficiency in manufacturing, if better/cheaper products are available and for diversification to other products. They contribute towards the following:

- Expansion, modernisation, and diversification for new products of existing plant facilities and operations.
- Product improvements to extend its life cycle.

1.24 Revival of Old Plants

It can be sometimes cheaper and faster to revive an idle/old plant instead of setting up an entirely new unit, since considerable infrastructure such as water storages, electrical installations, product storage tanks, and weighbridges would be already in place, or if most of the existing equipments are found in good condition.

However, the reasons for the plant being idle shall be thoroughly investigated and corrective steps planned carefully before investing in an attempt for revival.

1.25 Project Division for Expansion and Diversification

Instead of obtaining process know-how and technology from outside consultants or vendors, the company can become supplier of know-how developed in-house, provide consultancy to outside parties, and supply complete projects by creating a project division (may be in future) of their own.

The Research and Development division can be used to assist the project division for this. Some items of the plant and machinery can be manufactured through in-house fabrication facilities and procuring the rest externally for supply of the project to outside parties.

1.26 Cost Reduction

Management shall look into all unnecessary expenses and carry out cost-saving measures such as avoiding frequent change over to different products, using cheaper raw materials if possible, prolonging productive runs, and minimising shutdown periods. Unnecessary handling of materials, and wastage of power, water, and packaging items shall be avoided. Cogeneration shall be carried out by optimising process and steam conditions.

Chapter 2
Financial and Commercial Matters

2.1 Financial Costs

The following expenses are likely to be incurred for setting up a new unit:

- Appointment of a project team with technical, commercial, and administrative members,
- Expenses for market survey for shortlisting of products to be considered for manufacture,
- Appointment of consultants for getting basic engineering document (the fees will be about 2.5–4.0 % of project cost). The decision to go ahead can be taken if the BED indicates technical and financial feasibility,
- The consultants can now be requested to provide the detailed engineering document. Their fees could be about 5.0–10.0 % of project cost depending on the complexity of the technology (*please also see below*),
- Applying for various statutory licences and permissions for environmental clearance,
- Acquiring land, its levelling, fencing, and developing approach roads,
- Water supply and power connections (for construction as well as for running the plant),
- Further activities after a study of the *detailed engineering document* by the technical and financial teams are (i) floating inquiries for equipments and (ii) identifying suitable vendors (*on the basis of their technical knowledge, experience, and reliability*),
- Technical team members of the promoter may also wish to observe the actual working of similar equipments supplied by the vendor. The following activities are quite time-consuming and also involve costs to the promoters because the vendor's workshops may have to be visited to assess their fabrication capability:

K.R. Golwalkar, *Production Management of Chemical Industries*,
DOI 10.1007/978-3-319-28253-4_2

- Obtaining offers and shortlisting of vendors,
- Technical and commercial discussions with shortlisted vendors,
- Placing orders on finally selected vendor(s).

Advance amounts required by vendors are generally 15–25 % of cost of machinery (in case the cost of machinery is in the range of about 50–70 % of total cost of project depending on the complexity of technology). Vendors may ask for greater advance amount if costly materials of construction (stainless steel, titanium, or special alloys) are required for fabrication, pressure vessels are to be supplied or equipment designs are complicated such as special reactors or boilers.

2.2 Cost of Procurement

This needs additional funds for the following:

- Stage and final inspections at vendor's shop/at project site if it is a big item,
- Payment of further instalments to vendors/fabricators before it leaves fabrication shop,
- Packing, loading, forwarding, transit insurance, shipping (transport to site), unloading at site, storing, and site inspection again on receipt.

2.3 Cash Flow Planning for Site Jobs

In the meanwhile, the following should be ready at the project site:

- Establishment of site infrastructure—project office, stores, security arrangements, and other facilities such as transport, approach roads, and living quarters for site staff.
- Levelling of land and civil foundations as per design of the layout.
- Placing of equipments on the foundations can now be done. Care should be taken to correctly orient all nozzles—*this can need heavy-duty cranes and skilled manpower*. These machineries are hired at heavy charges on hourly basis.
- There are many major jobs such as fabrication of big storage tanks and large-diameter process vessels, internal refractory/acid-resistant brick lining and their curing, electrical cable work, fixing electrical motors to the respective equipments, installing internals in process units such as distribution trays, partition plates, loading of catalysts in converter, tower packing in gas absorption towers.

- These are followed by balance jobs such as fabricating/fixing prefabricated supports for ducts and pipelines, fabricating and then connecting ducts and piping to process vessels, installation of instrumentation probes and connecting cables, thermal insulation of process vessels and ducts/pipes.

2.3.1 Testing of Static Equipments

- Filling up vessels by water to check for any leak,
- Pressure tests by compressed air or hydraulic pressure tests by water (boiler, steam coils in evaporators, air receivers, and pressure vessels to be used for process),
- Smooth operation of valves, safety devices, and level indicators,
- Purging of dust and pipe scales,
- Checking the heating and cooling systems.

2.3.2 Testing and Mechanical Trial Runs of Moving/Rotating Machineries

- Air blowers and compressors,
- All pumps,
- Crushing and grinding mills,
- Belt conveyors, bucket elevators, and screw feeders,
- Agitated reactors, electrical cranes, and cooling tower fans, etc.

2.3.3 Statutory Inspections

These are necessary prior to commissioning of the plant and are to be done as per procedure instructed by the inspecting authorities. The following equipments are to be necessarily tested. Fees are to be paid to the statutory authorities for such inspection. This list is only suggestive, and more items can be added by them.

Pressure vessels, boilers and steam lines, fuel and gas lines, electrical equipments, and plant layout are inspected by Factory Inspectors from safety point of view.

2.3.4 Pollution Control Facilities and Waste Disposal Arrangements

Inspection of the facilities for treatment of liquid effluents and gaseous emissions; and disposal plants (with air pollution control) for solid wastes is carried out by State Pollution Control Authorities before granting consent to operate.

2.4 Cash Flow Planning

2.4.1 Typical Start-up Expenses

- Higher charges are to be paid for construction power till regular production starts (*please see more details in chapter on electrical installations*),
- Cost of obtaining fuels—making necessary safety enclosures and other arrangements for fulfilling statutory requirements for storage of fuels, transport to site, and unloading into storage tanks,
- Arranging for testing, transport, and storage of sufficient quantities of raw materials,
- Chemicals, lime, and alkali for the treatment of effluents,
- Some more typical items required for starting a chemical plant are—Filter cloth pieces, active carbon, storage of raw and treated water, starting lots of materials like sulphuric acids *and oleums for establishing initial circulations in absorption towers.*
- To keep additional spare parts and lubrication oils ready,
- *Fees to erection and commissioning engineers and their teams may be included in orders for equipments,*
- Additional labour will have to be employed (for the duration) till plant is commissioned and performance guarantee terms are fully met.

2.4.2 Sale of Products

It is generally seen that products from the trial runs and from the first few days of plant operations are not as per specifications given by clients. These cannot be supplied to them or sold in open markets. The initial quantities of such product need to be recycled/purified to make them saleable quality or disposed off at very low price to some party who may be able to use them.

During such times, very little revenue is obtained, while money is being spent continuously for getting raw materials and other necessary inputs. **The cash flow must be planned for such situations also**.

Blackish sulphuric acid with some suspended matter may be produced in a new plant for first few days due to the dust and scales in process units and pipelines even after they are flushed out. This acid can be sold to manufacturer of single superphosphate or alum (to be used for effluent treatment).

2.5 Investment of Own Funds by the Promoters

These could be up to 20 % of project cost before financial institutes agree to release loans. There will be certain processing fees (0.5–1.0 % of loan amount) charged by the financial institutes. Certain other expenses will also be there for:

- Mortgaging immovable properties for obtaining loans,
- Legal documentation and fees by legal consultants.

Venture capital may be obtained if it is a novel technology or a scheme with high technology. This investment provides the seed capital or an initial fund for getting further capital from financial institutes. It is for new companies which are not well known, have limited operations earlier or are too small to raise capital in the public markets and may not be able to get bank loans or raise capital through public offering for shares.

Venture capitalists usually try to get significant control over company decisions and a large share of the company's ownership because the parties giving venture capital are taking a higher risk of business failure in case the project fails (since the investments are being made in smaller and less known companies—*who may not be able to generate the expected return on the investment*).

- Raising funds from public as equity shares, preference shares, and debentures.
- Getting loans from financial institutions and banks (government/private)—the promoters are generally required to invest their own funds to the extent of 20–25 % of the project cost as seed money.

Cost of production per unit will be high till plant operations are stabilised and rated capacity is reached. Fine-tuning of operations is to be carried out carefully to establish all procedures of material feeding, starting the plant, operating various controls, and maintenance procedures. The cost of production per unit output will start coming down after this.

2.6 Hooking up with Running Plant

When new equipments are to be installed in a running plant, most of the infrastructure is already in place; however, very careful planning is required to erect the new equipments along with accessories (some of which can be supplied by the vendor, some may have to be arranged by the purchaser, while certain items may be already available in the premises). It is advisable to take mechanical trials of as many as possible new machineries and complete the inspections of new equipments before hooking up to the old plant.

2.7 Depreciation

Process equipments and machineries in a chemical plant are subjected to corrosive conditions. Protective linings and external paints are generally provided, but these do get damaged earlier as compared to units in other industries.

Replacements become necessary sooner as compared to the units in mechanical workshops or civil structures for offices/residential buildings.

It is a thumb rule that almost all the reaction units and machines handling chemicals will have to be replaced every five years as they would be quite corroded or become unserviceable. Hence, managements shall keep aside an amount equal to at least 20 % of their costs every year.

This is a practical allocation from the profits and is generally allowed as a legitimate expense for running the industry by government taxation departments of many countries. The taxable income of the industry is reduced by this amount accordingly.

Efforts shall, however, be made to increase the useful life of the items by a few more years. This can create little surplus funds which shall be carried forward to the next financial year to meet any sudden major breakdown or some unforeseen mishap.

2.8 Government Assistance and Tax Concessions

These are given by certain countries:

- For establishing the industry in special geographical areas where more unemployment may be prevalent in the local population,
- In certain specially established export promotion zones where the government wants to promote export business,
- For certain research activities,
- Lower rate of taxation or deferred taxation (applicable taxes to be paid later on).

However, the feasibility and satisfactory return on investment shall be independently established even without such government assistance.

2.9 Marketing of Products

2.9.1 Production for Export/Prestigious Clients

The sale of products to prestigious clients or in export markets can create good reputation for the organisation and generate more business in future. It is very necessary that the goodwill must be maintained for further progress. The intended use of the product must be clearly understood by discussing with the client so that special precautions can be taken while procuring the raw materials and packing in suitable sized containers.

2.9.2 Selling the Product

Free samples of the product or trial runs with the products may be offered to the client for generating confidence. Some of the products could be pour point depressant for oils, anti-caking agent for salt, and descaling agents for heat exchanger tubes.

The selling price may include the loading and transport charges, cost of containers, taxes and other levies, and insurance charges (the price then becomes the *delivered cost at site* to client).

Sale on credit—clients may be asked to pay only partly for the consignment and allow payment of balance amount after confirming that the product was found useful indeed.

Methyl violet dye can be sold as crystals which need to be weighed and dissolved or as a solution in suitable solvent which is easier to handle and feed in controlled doses at the point of use.

Papermaker's alum can be sold as solid or as 8–10 % solution (the latter is easy to use).

2.9.3 Instruction Manuals

- Instructions for unloading and opening the containers (e.g. *do not use sharp tools or hammers for opening*),
- Instructions shall be given how to properly wash and clean the drums/small cans before returning in case the drums (100/200 litres capacity) or smaller cans (10/20/50 litres capacity) are supplied on a returnable basis (in order to save cost to the clients)—*no water shall remain inside when they are returned,*
- In case the drums are not to be returned, advice shall be made available for their safe disposal,
- Instructions for correctly and safely using the product:

How to handle the cans or drums, how to dissolve the product, and how to add slowly to the process unit at point of use while standing at a place not to inhale the vapour/dust.

Visits may be undertaken to explain the above matters for creating more goodwill and to answer any queries by clients on the spot.

The essential matters which should be taken care of/be addressed are as follows.

2.9.4 Confirmation of Product Quality

All batches/lots being sold must be thoroughly checked by quality control laboratory in order to confirm that the specifications given by clients are fully met or even exceeded (the product can be even better than asked for).

The minimum purity level must be met by all consignments. In addition, the maximum level of impurities shall never exceed the limits set in the order.

A particular physical form is sometimes required by the client such as:

- Absence of big lumps—*since these do not dissolve quickly,*
- Requirement as a free flowing powder,
- Granules of a particular size or coated with anti-caking agents.

The test certificates for such analysis should mention the reference of order (name of client, order number, a copy of original specifications), production batch number, date of manufacture, and manufacturer's code—*for ease of identification of the reaction units, process plants, operation sequence, or conditions maintained during production on the date of manufacture in future. These will be useful if any communication (complaint or compliment) is received about the product.*

Taking care of quality—*Raw materials shall be thoroughly checked before using them for manufacture. The process streams shall also be checked at all stages of manufacture such as after calcinations or filtration of raw materials before charging into reactors; at exit of reactors, after every purification stage (settling–filtration–distillation) so that any process problems can be pinpointed and corrected.*

2.9.5 Internal Records (For Manufacturer)

Internal records (for manufacturer) are as follows—date and shift of manufacture, reactor/unit used for production, process conditions maintained during production, and source and analysis of raw materials used.

The following are the documents to be provided while dispatching the products:

- Containers must be correctly labelled with name and analysis of the product.
- Material safety data sheets giving all important properties such as boiling and melting points, miscibility with water and other solvents, inflammable nature, toxicity, precautions to be taken for safe handling, and how to collect and make a spillage harmless (absorb in ash/washing by copious amounts of water).
- If there is a possibility of material spillage due to wrong handling, write antidotes and steps for first aid prominently on the packages in as many languages as possible with at least two internationally known ones. This is to suggest remedial action if material is accidentally inhaled/spilled on body.
- Precautions for handling glass containers shall also be written prominently on the packages in as many languages as possible with at least two internationally known ones. Stickers with standard symbols shall also be fixed, and symbols shall be printed.
- Information should also be given if the product is not compatible with any other material so that clients can take proper precaution during storage (not exposing to sunlight or rains) and using it (specially the feeding sequence to process reactors).
- Antidotes/first aid to be given to affected persons in case of a spillage must be prominently mentioned on every package/container.
- Confirmation of quantity packed in each container.
- Size of packages shall be as required by clients: these should be easy to fill, load, shift, unload, store, and open for use. The tare weight, net contents, and total weight of the packages shall be clearly marked on each package and also mentioned in the documents.

  Small bags carrying sulphur powder generally have a net content of 50 kg material and a tare weight of 1.0–1.5 kg; thus, the total weight of the bags may be 51.0–51.5 kg.

- Some precautions may be specified by the authorities for transport by railway tankers or road tankers before the consignments are accepted for carrying them. It may be necessary to pay higher insurance charges and declare all details about the products if required by statutory authorities prior to export consignments.

2.9.6 Dispatch of Products and Storing at Client's Site

2.9.6.1 Drop Tests and Stack Tests

Drop tests and stack tests must be carried out on a sufficiently large number of carbuoys if they are to be used for carrying dangerous liquids. These tests are done by dropping a filled-up carbuoy from a known height—say from 3 to 4 m, to confirm that they will not leak during shipment. Likewise, stacks of carbuoys are arranged one above the other to confirm that the lowest one will not buckle under the load and start leaking.

The requirements are to be worked out by the manufacturer of the product who is in a better position to take such trials in the plant before dispatch.

Provide lifting lugs or slings to the containers for loading to transport vehicles. This is to prevent the use of hooks or any sharp tools for lifting the bags/containers which can tear them and cause spillage.

2.9.6.2 Seaworthy Containers

Seaworthy containers shall be used for shipments by sea since they may get corroded by salty ambient conditions in case they are carried on decks. These shall be provided safe enclosures with soft packing around and then finally packed in a strong container.

Special precautions are to be taken for packaging when transport of high-value materials is by air (samples/special chemicals/pharmaceuticals/perfumery products).

Advice shall be given whether the material (product) can be moved in the premises by belt conveyors/bucket elevators or screw feeders, etc. *It will depend on inflammable/toxic/hygroscopic properties and angle of repose of the material.*

2.9.6.3 Storage Precautions

Storage precautions for storing the containers at the user end are to be given. Some of the precautions to be avoided are direct sunlight, exposure to rains, dusty conditions, proximity to hot surfaces, provision of earth connection to container (having aluminium powder, inflammable matter), and need to install smoke detectors and water sprinklers in the storage sheds.

Whether the material can be stored in bins or silos with a rotary discharge valve at the bottom?

These precautions depend on the particular product, and clients shall be made familiar with them by giving training for handling the products if necessary.

2.9.7 Delivery Schedules

The delivery schedules shall be committed to the clients only after consulting the plant engineers who are operating and maintaining the production units, because some of the process units such as reactors, condensers, and filters may not be working to rated capacity or a plant shutdown might have been planned for their repairs or cleaning.

Therefore, the copies of delivery orders (finally accepted for supply to clients) shall be given to departments responsible for purchase of raw materials and other necessary items, production and maintenance staff, transport section, etc., to enable them planning of their activities.

The following situations can affect the timely delivery of products:

- Delay in procurement of the specified raw materials, catalysts, maintenance spares, and packing items due to any reason,
- Adverse weather conditions prevailing at present (heavy continuous rains can affect drying operations),
- Scheduled preventive programmes planned for process units,
- Any urgent repairs to be carried out to key process units,
- Statutory inspection of some key equipments has become due, and it is necessary to stop the plant, overhaul and repair (*wherever necessary*) and present to authorities for annual inspection as per the law, e.g. steam-generating boilers, pressure vessels, electrical installations, and safety devices.
- Shortage of water, power, fuel, or transport bottlenecks
- In rare situations—absence of key personnel
- Finance managers may ask for certain conditions to be met before products are dispatched to an overseas client. Some of these could relate to the performance guarantee terms (these must be finalised in consultation with technical persons), payment of an advance amount by the client, credit terms asked for by client, whether client agrees to pay the costs of packing, insurance and transport for the consignments, and any other taxes and levies to be paid to statutory authorities in supplier's country and client's country.

2.9.8 A Word of Caution

The marketing department shall never insist the plant engineers to run the plant above the rated capacity in order to meet the commitments made to the clients for the total quantity and delivery dates.

This can force them to run the process units by occasionally overloading them—which can be dangerous and cause environmental pollution or inefficient operation of the plant.

In rare cases, the equipment life may get reduced (e.g. charging excessive materials in a batch reactor and then running agitators at too high speed, **or** operating at too high SO_2 % to converter for producing more oleums from a sulphuric acid plant).

2.10 Agreements for Technical Consultancy

It will be found useful to get thorough professional advice (both technical and commercial) while setting up a new project, for expansion of a running plant or for revival of an idle plant. Technical advice may also be required if some problems are not getting solved by own technical teams and a fresh look is required. Managements may also enter into agreements for process know-how or engineering designs and then proceed to procure or fabricate the equipments instead of setting up a project on a turnkey basis in order to save costs.

The agreements for consultancy could be made for the following:

- Market survey to assess demand for a product and its availability,
- Obtaining only process know-how about a new modernised version of the existing process or a completely new process and then proceeding further on their own,
- Basic engineering package to get more details about the process description, major equipments required, raw materials and energy consumptions, an approximate requirement of funds, area, manpower for the plant, return on investment, and payback period. The management can now decide whether to go ahead, keep in abeyance, or abandon the project,
- Detailed engineering package (DEP) shall include but will not be limited to the following:
  - Full details of the process, equipments, and process piping,
  - Material and energy balances,
  - Details of safety devices and safety interconnections,
  - Electrical load data and wiring diagrams,
  - Civil and architectural drawings,
  - Effluent generation and treatment thereof by pollution control schemes,
  - Assistance for procurement (preparation of inquiry documents, evaluation of offers received from vendors, shortlisting, and final selection of vendors on technical basis), commercial discussions, and shortlisting are generally done by managements themselves,
  - Erection and commissioning of the plant equipments and machineries and reaching rated output from the plant,
  - The performance guarantee test runs are the most important part of such agreements as they represent completion of the project and the handing over to the regular operational team. Guaranteed maximum raw material and utilities consumptions (will not exceed these figures per unit output of saleable quality of finished product),
  - Operation and maintenance manuals shall also form part of the DEP. These shall include dos and don'ts for the specific guidance of the shop floor personnel.

- Training may also be arranged on a running unit or given on the plant being erected and commissioned. This may be charged separately on a man-day basis by the consultants,
- The consultants shall reply to any specific queries from the clients or from statutory authorities regarding technical matters, safety issues, effluent generation, and treatment thereof.

- Expansion of a running production plant:
  It is necessary to look into all the process units and the site infrastructure, including electrical installations, water supply, storage tanks for raw materials and products, and the handling equipments and systems. The consultants should examine all these very carefully and check suitability of the facilities for increased production (as per product mix decided). Specifically, the capacity of effluent treatment plant, main electrical transformers, refrigeration plants, cooling towers, fuel tanks, etc., shall be checked. All bottlenecks shall be identified and upgraded. It may become necessary to procure equipments/machineries with higher capacity wherever existing units cannot be upgraded.

Special care shall be taken to confirm that no pressure vessel will operate beyond its permitted pressure or electrical equipments (including distribution network in the premises) will get overloaded when the plant capacity is expanded. Moreover, no unsafe condition shall be created anywhere in the plant due to expansion of the capacity.

- Diversification of a running production plant:
  In this case, the agreement with consultants shall include a clause that as much of the existing process units and machineries will be used for the manufacture of additional (new) products. The material of construction shall be carefully looked into so that there will be no contamination and corrosion issues for the new products. Only a minimum of additional units and machineries shall be added, and this shall be done with a minimum interruption of production run of the plant. As before, the capacity of effluent treatment plant, main electrical transformers, refrigeration plants, cooling towers, fuel tanks, etc., shall be specially checked.

No unsafe condition shall be created anywhere in the plant due to diversification of the plant to new products.

A performance guarantee test shall be included in the contract with the consultants.

2.10.1 General Conditions

It is advisable to enter into a mutual secrecy agreement so that neither the consultants nor the plant management will disclose any of the plant design, operating data, or any other information to outside parties.

There shall be an agreement for regular visits to the plant by the consultants—say one visit per month for the first six months and once in three months thereafter. The first two or three visits shall be free of cost (except travel fare and arrangement of stay at site to be made by the plant authorities) to the consultants. All subsequent visits may be on a chargeable basis of per man-day. This will be convenient to both parties.

Any other conditions may be included in the contract after mutual agreement.

2.11 Running the Plant on Contract Basis

It may be found convenient or even economical in certain circumstances to run the plant activities (either wholly or partially) by outside parties on contract.

Some of these circumstances could be when the purchaser does not have:

- Enough funds,
- Proper and sufficient manpower,
- Process and engineering know-how,
- Identify a project which will be profitable or is under statutory instructions to set up such a unit (e.g. effluent treatment plant),
- The purchaser has plans to raise funds in future to take over such a plant and is willing to pay a higher price for the same because the working has been established in the meanwhile.

2.11.1 Typical Options for External Assistance

- BOT: Build, Operate, and Transfer. The plant is built and operated by external agency till performance guarantee terms are met and then transferred immediately to the purchaser. The purchaser thus gets a ready-made new running plant (e.g. effluent treatment).
- BOMT: Build, Operate, Maintain, and Transfer to purchaser. Major control on the plant remains with the external agency till it is transferred. Maintenance of the plant is also done by the agency. At the end of the contracted period—*which should not be too long*—it is to be handed over to purchaser. But it must be maintained in a good condition.

- BOOM: Build, Own, Operate, and Maintain for some years as per agreement between purchaser and external agency. The profits during these years could be shared or taken by external agency. It is then to be transferred to the original party for an amount (as per agreement made earlier) who has given the contract. Major control of the plant will remain with the external agency for the contracted period.

The terms and conditions for sharing of profits and costs shall be worked out and written down in the agreements accordingly.

2.11.2 Agreement Between the Purchaser and External Party

- The plant should be safely operable in the local geographical and climatic condition (ambient temperatures, rainfall, etc.) with the technology offered. Locally available (if cheaper) raw materials shall be used to the maximum extent in the plant. This shall be discussed between the purchaser and the contractor (external party) and confirmed in the agreement. Dependence on imported items shall be minimum.
- Consumption of raw materials, power, steam, and water shall be lowest or comparable with the best in industry. Specific consumptions of all inputs per unit of output shall be very economical over all operating capacity ranges. *This is the main requirement when the plant is being obtained on a turnkey basis from the external party.*
- A clause for operability of the plant in the range of 80 to 130 % of the capacity in an *efficient and safe manner* shall be written in the agreement. This is to safeguard when the demand for products is less (to run at 80 % capacity), or at overload of rated capacity (to run up to 130 %) if higher demand for product is to be met urgently.

2.11.3 Important Jobs to Be Done by Purchaser

These are to be done by purchaser themselves even after obtaining services of external agency on contract basis. Some of these are as follows:

- Obtaining all statutory permissions to establish and operate, and arranging land, power, water supply, and all necessary infrastructure. All statutory requirements and instructions shall be complied with by *the owner*.
- The purchaser shall appoint proper manpower who should be able to understand the process technology and able to operate and maintain it when it is handed over to them so that services of external agency are not required afterwards.

2.12 Organisational Resource Planning

This shall be done for achieving the goal of developing the organisation as an efficient, reliable manufacturer which delivers quality products as per specifications and delivery schedules committed, while operating safely without causing environmental pollution.

Software shall be developed in-house in consultation with all concerned departments to address the important activities as below. The software shall have built-in alert signals to draw the attention of senior personnel in each department if any activity is not meeting the targets set. This will ensure timely action for properly planning the organisational resources.

2.12.1 Fulfilling of All Orders Accepted by the Sales Department

This involves the tracking of the following:

- Names of clients,
- Names of products and their specifications,
- Delivery schedules accepted,
- Availability in the storage tanks/sheds of these products (with certified quality),
- Approvals by clients for the samples given to them,
- Arrangements for dispatch and confirming acceptance by clients—*functioning of weighbridges, availability of special vehicles for transport of dangerous products, any transport bottlenecks, confirming receipt by clients, and payment of bills by clients.*

2.12.2 Tracking All Purchase Orders Issued

Track all purchase orders issued to suppliers of raw materials and inputs—approvals of samples received given by suppliers, any transport bottlenecks, actual quantities received, bills raised by vendors, payments actually made, balance payment amounts, and proper storage at site.

2.12.3 Manufacturing/Production Operations

- Confirming capacity of process units and reactors for each type of product to be manufactured and their actual availability for the production by discussing with senior production and maintenance engineers,

- Confirming availability of required quantities of all raw materials and inputs including additives, property modifiers, and packing containers for the products,
- Availability of water, power, and fuel as per amounts required,
- Tracking quantities of all items actually issued from stores and their balance stocks,
- Amounts of various products actually produced,
- Quality control to confirm the quantities of finished items which meet required specifications,
- **Statutory inspections** due in the meanwhile which may need stopping some of the units (specially pressure vessels).

2.12.4 Safety Issues

Safety issues to be addressed urgently

- If there are chances of accidents when production rates *may have to be increased* beyond certain limits for fulfilling urgent orders from clients,
- Some minor leaks of acids/gases from reactor are yet to be attended,
- Some pending maintenance job (e.g. a support leg of process reactor has become weak and needs to be immediately reinforced).

2.12.5 Effluent Treatment

Capacity of effluent treatment plant shall be confirmed to handle the effluent load generated by the product mix if extra amounts of effluents are likely to get generated. Temporary storage of additional effluents may be provided for properly treating them later on before discharge.

2.12.6 Maintenance

Any plant stoppage is planned for carrying out preventive maintenance.

Any urgent equipment repair/replacement programme is planned.

2.12.7 Effect of Weather Conditions

Drying operations may get affected or the dried product may absorb moisture if it is raining heavily or humidity is high.

Transport of highly volatile/inflammable products needs more precautions during severe summer.

There is possibility of pipelines of process plants or outlet valves of tankers getting jammed or choked due to solidification in severe winter.

2.12.8 Manpower

Holiday season will have possibility of absenteeism or key personnel already gone on leave.

2.13 Resource Allocations for Increasing Revenue

Resources are of various types such as installed machineries, infrastructure facilities, trained manpower, and cash funds. No resource shall remain idle as far as possible. It is suggested to consider how all of them can be utilised efficiently.

- Financial—cash, credit facilities, and advances paid to vendors,
- Manpower—qualified, experienced engineers and technicians who can be sent on deputation or for providing consultancy services to (similar) industries,
- Equipments installed in the premises which can be removed from position and sent outside, e.g. crushers, small packaged boilers, dissolvers, and centrifuges
- Spare capacity available in the installed equipments (which are fixed on position) for taking up outside jobs,
- Excess power, steam, hot water, hot air generated available from waste heat recovery units, and cogeneration facilities,
- Maintenance facilities and fabrication machineries,
- Laboratory facilities which can take up outside jobs,
- Materials in stock (excess quantity can be given on loan or sold),
- Utilities and infrastructure (water treatment plants, weighbridges, warehouse, transport fleet).

2.13.1 Cost of Idle Resources to the Organisation

Investment is already made in acquiring various resources such as in production units, auxiliary equipments, and utilities by the organisation for certain rates of production. The unit cost of product is optimum when the plant is run at its rated capacity. However, if the plant is operated below its rated capacity, the consequences can be:

- Some processing units and machinery can become idle or redundant (the capital investment made for them remains unutilised; in addition, they have to be preserved for use in future and insurance premiums are to be paid to insurance companies). These equipments may even start getting rusted or jammed if they are kept stopped for long time.
- Power and water consumption per unit of output can be excessive when bigger machines such as crushers, big agitated reactors, boilers, pumps, and blowers are run much *below the lowest recommended point of operation by the manufacturer.*
- Trained manpower can become idle if they are already employed on a permanent basis or engaged on long-term contract if there is no work for them.
- If spare parts are already procured (corresponding to running of the plant at rated capacity), some of the spares can get rusted/jammed or obsolete due to long storage.
- Packing, bag stitching, and weighing machinery are underutilised.
- The transport fleet (which may consist of specially designed vehicles) becomes idle.
- Captive power generation or cogeneration facilities may have to be operated below their optimum efficiency point or may become partially idle.
- However, cost of purchase and taxes have already been incurred/paid on obtaining and installing these equipments (*but they may not be required now*) while licensing fees, royalties for process know-how, and catalysts charged in the reactors are to be paid on a recurring basis. The erection and commissioning expenses for the equipment have already been incurred.
- Various other inputs (power and water supply) have been arranged at considerable costs. In addition, the interest payment on borrowed funds becomes due as per repayment schedules (generally as monthly instalments).

All such expenses are now distributed over a *lower output* from the plant. This can increase the cost of production per unit and can be recovered only by raising the selling price.

2.13.2 Some Typical Reasons for Resources Becoming Idle

- There is not enough demand for the products (cheaper, better product has become available from competitors).
- Raw materials of required specifications are not available while the raw materials which are available need considerable purifications, drying, etc.
- Some key process unit or machinery is not working properly or stopped for maintenance for long time.
- Catalyst has got deactivated, and fresh supplies are not immediately available.
- Water supply is not available due to less rainfall and less storage in main supply reservoir.

- Fuel supply is delayed.
- External power supply lines have broken down.
- Transport fleet for outgoing products is under maintenance, or the road is blocked because some bridge has broken down.
- There are some pending demands from workforce.
- Plant is generating more effluent, and objections are raised by authorities/surrounding population forcing curtailment of production rates.
- Statutory authorities have instructed to stop the plant for some urgent corrective actions or inspections—**alert management should never allow such a thing to happen.**

2.13.3 Remedial Actions

Some corrective steps appear obvious and quite simple—but there could be practical difficulties in their implementation. Everyone in the organisation shall make sincere efforts to overcome the situation.

As a first step, all matters related to safety of personnel and plant machinery, environmental pollution control, and instructions from statutory authorities shall be attended immediately.

A programme for continuous innovation, suggestion schemes, etc., shall be in operation for improvement in the product and to bring down the cost of production. Condition monitoring of all key units and machineries and preventive maintenance will be able to minimise long stoppages of the plant. There shall be regular coordination meetings among operation and maintenance engineers to analyse reasons of breakdowns and to develop better methods of operation as well as more robust designs of equipments and machineries.

Auxiliary equipments for pretreatment of raw materials shall be in place to prevent deactivation of catalyst and fouling of heat transfer surfaces. Formats have been given elsewhere in this book for drawing attention to key equipments and consequences of their breakdown.

Stocks of all necessary raw materials and other inputs shall be monitored on a daily basis. Production planning engineers must know every day how long the available stocks will last to run the plant as per (current rates of production) the product mix. The reorder point for replenishment (the lowest stock level) of these items shall be clearly known with automatic warning available.

Grievance cells and other mechanisms for resolution of conflicts with workforce shall always be in force. No legitimate problem shall be ignored. Officers of this department as well as other senior engineers and managerial staff shall be alert to see that any discontent brewing in the organisation does not snowball into a major conflict.

Alternate arrangements for transport shall be made in advance in anticipation of heavy rains or for situations when there are chances of interruption. Managements may also look into the possibility to build up stocks of their products at the customer's plants or in warehouses at a convenient location near to the customer.

2.13.4 Allocation of Resources

Allocation from existing resources for new products and expansion/modernisation activities shall be done *only after observing that surplus* power, water, storage space, manpower, facilities for maintenance and fabrication in-house, and cash flow are available. This can be done by examining these resources in detail on the following basis. The list is only suggestive, and more criteria can be added for *establishing that these are surplus resources indeed.*

- The rated capacity of certain equipment/machine is more than necessary.
- The actual output from such machines/process units is also more than necessary when it is run all the time. Hence, it is required to run it only for some time in a shift.
- Required running time of the machine for processing the input is only a few hours per shift/day without disturbing the main plant being run at rated capacity. Hence, it is to be checked that can it be spared for a few hours every shift/day?
- Does this machine/resource have multiple uses (grinding raw material for feeding to many reactors/supplying steam to many reactors)?
- Raw material stock—Is this raw material used for making different products? In that case, the raw material shall be used for manufacture of products which will give higher profits or which are in stock in a very short quantity and an important customer has placed urgent orders. The raw material can also be used for a product where the reactor is in a ready condition for charging, while other reactors are undergoing maintenance or are being cleaned. Plant manager shall decide in consultation with marketing and maintenance departments and use the excess raw material in stock accordingly. Any further surplus may be sold outside.
- How much treated water can be generated per day and how much is actually required? Can the treated water be sold outside?
- How much steam is generated per hour and how much is required in the plant? Is a nearby customer willing to buy the surplus steam? Can some outside party be allowed to set up an evaporator in the premises for concentrating their dilute solution/to set up a dryer for drying wet material and then charged for the steam consumed.
- The steam consumption can be determined by collecting the condensate and measuring it. The condensate can be recycled to save cost of fresh DM water *provided it is not contaminated.*

- What is the total power required to run the plant at its rated capacity and how much power is getting generated through the steam turbogenerator? How much is the surplus? How much steam can be exported if cogeneration system is installed in the plant?
- Financial—Invest the surplus cash in important inputs/buying good-quality material in larger amount if available at off-season discount, and innovation and research activities.
- Procure items on credit if possible.
- Equipments installed in the premises which can be removed from position and sent outside, e.g. small crushers, small packaged boilers, dissolvers, and centrifuges if they are used sparingly. They may be used for two/three days continuously to build up stock of materials and then lent on hire.

2.13.5 Spare Capacity Available

- Taking up outside jobs by installed equipments (which are fixed on position)—crushers, grinders, dryers, and filters can be used to work for outside parties.
- Maintenance facilities can be used for repair or fabrication work for outside parties.
- Laboratory facilities can be used for certification work for outside parties.
- Outside parties can be allowed to use weighbridges, warehouse, and transport fleet for certain fees, and thus, some revenue can be generated.
- Are trained engineers and technicians available for sending them to assignments outside the premises (to other units for providing consultancy services on a per man-day chargeable basis).

2.14 Cost Reduction

Records of total production and plant stoppages on a daily and monthly basis shall be maintained. Similarly, records of consumptions of raw materials, power, steam, water, and spare parts shall also be available. These shall be updated regularly for getting the latest figures at a glance.

2.14.1 Factors Contributing to Variable Cost of Production

- The above data shall be studied, and the quantity of acceptable grade products manufactured shall be determined by subtracting the quantity of rejected materials from the total production figures. The quantity of rejected products (which may be reprocessed and sold thereafter) shall also be subtracted because they need additional inputs for making them acceptable in market. Consumptions of inputs shall be calculated on the basis of only marketable grade products.

- Generate the data for cost of consumptions of raw materials, power, steam, water, and spare parts on the above basis. Consider only the higher of the costs in the period under study *because some inputs might have become cheaper meanwhile*. Select the items for further detailed study which contribute about 80 % of the total costs. Remaining items are also to be looked into after corrective steps are taken for the above.
- Where same facility (e.g. grinding units) or inputs (e.g. steam supply) are used/consumed, allocate the consumptions in proportion of quantities of each item produced and quantity of effluent load attributed to such items.
- Compare landed cost of all raw materials at point of charging into reaction units —since *apparently* cheaper raw materials need lot of pretreatment such as filtration, drying, and screening before charging into process units. The procurement can be seasonal and in economic lots if lower rates or credit facilities are available.

2.14.2 Some Tips for Reducing Variable Cost

- If possible, direct unloading into process units shall be explored to save cost and contamination due to repeated handling in the premises (sulphur procured in road trucks can be unloaded directly in the melter, rock phosphate unloaded directly on to the grids of crushers, bauxite lumps into loading hoppers of belt conveyors, and alkali in day tanks) **or**
- Buy small bags or packets of additives which can be directly charged into reactors.
- Select the required fuel on the basis of *actually available heat* in Kcal per unit cost instead of gross calorific value and cost per unit weight.
- Use only treated water for cooling, making solutions, as aqueous medium for process reactions. Use spent water from one unit for next unit—blowdowns from cooling tower for flushing pipes, washing floors, gardening, cleaning drains, and basins.
- Compare the actual yield with expected yield. Investigate the loss of materials through effluent streams. Analyse samples of wash liquors from process units and spent scrubbing system liquors to check whether any product or reactants are getting lost.
- Examine weighbridges and recalibrate by using standard known weights in case of doubt.

- Sell the products in returnable packing/containers of bigger sizes (100–200 l instead of 5–10 l) to reduce workforce and time required for filling such small containers.
- Agitated reactors: Provide simple removable baffles of different designs. Run agitators at lower speed or less running time. Trials can now be taken to minimise power consumption.
- Maximise heat recovery from all possible sources. Use outgoing hot process streams for preheating of incoming cold process streams. (*Outgoing depleted hot oleum stream at 125–130 °C from oleum boiler is used to preheat the incoming concentrated oleum stream at 50–55 °C into the oleum boiler by an oleum/oleum heat exchanger in a liquid SO_3 plant*).
- Avoid unnecessary high pressure for the generation of steam unless cogeneration is to be done. Since the saturation temperature of high-pressure steam is high, the flue gases can exit at high temperature from the boiler—few degrees above the saturation temperature. This may cause loss of heat unless additional heat recovery units are provided.
- Provide variable-frequency drive motors for pumps and blowers wherever possible instead of throttling the delivery valves for flow control.
- Reuse spent scrubbing liquor for the regeneration of alkali/floor washing/cleaning drains.
- Implement inventory control programme through standardisation of spares and detailed analysis of reasons for consumptions by joint consultation among chemical, mechanical, electrical teams and purchase officers from engineering stores.
- Automation will lead to rationalisation of workforce and reduce excess manpower.
- Internal audits shall be done regularly. Test the plant performance every 3–6 months as if it is the performance guarantee test. This will point out inefficient units in the plant.
- Own transport fleet may be used for reliable and safe transportation of special, costly, or dangerous materials.
- Use metering tanks for feeding raw materials and for selling products.
- Prolong production run of the plant to minimise the proportion of the start-up, stabilisation, and shutdown periods in the total cycle time as the production rate is less during these days.
- Depute sales teams to client's works to explain handling and using the products to minimise rejection of products. Arrange direct sales from works to clients to eliminate middlemen.

2.15 Total Quality Management (TQM)

Progressive managements follow a total quality management (TQM) approach to build up reputation of the organisation for consistent good-quality products and credibility for meeting the delivery schedules. TQM indicates the culture, attitude, and organisation of the company which is making a serious continuous effort to provide products as per specifications given by the customers.

It shows the organisation approach that quality is the main goal towards which the activities for design, planning, and improvement are directed.

2.15.1 Quality and Other Goals

However, apart from quality, the goals of the chemical industry shall be safety, pollution control, energy efficiency, product quality, and to maintain rated production capacity with improvement in market share and customer satisfaction which is *reflected in repeat orders for the products.*

All plant engineers must therefore pay adequate attention to the following matters: safety, maintenance, pollution control, careful operation of the plant and machinery (which can increase equipment life), innovation, research and development, compliance with all statutory instructions and regulations, correct specifications for raw material, and inputs as these can contribute significantly to meet the goals of the company.

The culture requires quality in all processes and operations (which shall be done according to standard operating procedures) *right at the first time itself* while making maximum efforts to meet the above goals and eliminate off-spec products and minimise the generation of effluents.

2.15.2 The Key Elements for Successful Implementation of TQM

Leadership by the top management is most important. They shall lead by personal involvement to set example and shall inspire all members of the organisation (seniors and juniors) to meet the goals of the company towards TQM approach and should instil values for safety, pollution control, and quality. The leadership shall provide guidelines and directions to subordinates that are understood by all and create well-defined systems, as well as methods for their implementation and performance measures for achieving those goals.

Targets shall be set to operate the plants without any accident, while always meeting norms set by state authorities for effluent characteristics and generation, and meeting specifications for the products. Standard operating procedures shall be well defined and explained to all concerned. Important parameters shall be displayed prominently on notice boards and as standing instructions. Any deviations shall be immediately investigated and corrective actions carried out (e.g. repair to safety valve; to scrubbing system; to process control instrumentation).

The top leadership, seniors, and juniors shall work as a team for success of TQM. The seniors must understand, believe, and practise TQM. They should make sure that strategies, values, and goals are well communicated throughout the organisation to promote focus and clarity.

Management shall create an environment where there is no fear to share knowledge (which is obtained from experience in the production plant) and suggestions for innovations. All team members shall be motivated to trust each other—due to ethical work culture.

The seniors must give credit where credit is due. This can further motivate juniors to carry out their jobs more sincerely and attentively and results in a successful TQM organisation.

Ethical working: Code of conduct (guidelines) that all employees should follow while carrying out their work. If a mistake is committed during operation by anyone, he/she should not hide it but try to correct it and inform coworking persons and seniors. It develops openness, fairness, sincerity, and involvement by everyone. *It will prevent recurrence of the same since the experience shall be shared with others.*

Integrity: Everyone in the organisation shall perform their tasks honestly and sincerely while being fair and adhering to the facts. *A customer will respect and place repeat orders on a company which has integrity in its workforce.*

TQM is built on **trust** which is generated by integrity and ethical conduct.

It promotes involvement and commitment of all members and hence the cooperation among them which is essential for TQM.

Team members shall trust the integrity, honesty, and sincerity of each other and decisions which are for continuous improvement of the process and products.

Communication: The success of TQM depends on clear communication among all the departments, seniors and juniors, and technical and non-technical personnel in the organisation and external suppliers and customers. Persons receiving the message must clearly understand the ideas the sender intended to communicate—when they are specially related to the required product quality and delivery schedules. Only correct information shall be communicated by the sender to the receiver to maintain the trust.

Example: Suppliers shall understand the need for urgent supply of items such as catalysts, spares for a process control instrument, and stabilisers for the finished product when a purchase order (*actually it will be an urgent communication*) is sent to them. They shall **act fast** because these items are required for ensuring proper reactions, and process and quality control, preventing deterioration of finished products.

Supervisors must listen carefully whereby employees can send about the TQM process. There are different channels of communication such as:

- The seniors shall make the junior employees understand clearly about TQM by providing guidelines and directions for meeting goals of their organisation. These shall be safety, pollution control, energy efficiency, product quality, maintaining rated production capacity, improvement in market share, and customer satisfaction (which is reflected in repeat orders for the products).
- The junior employees should be able to provide suggestions to seniors/higher management suggestions for improvement or inform genuine difficulties faced by them (during operations). Supervisors must listen to such constructive criticism, carefully and sympathetically. It can be useful to correct the situation that comes about. This forms a level of trust between supervisors and employees.

Communication between departments
There shall clearly understand each other's requirements and difficulties and should listen to suggestions given for improvement in working to make it easier. Marketing shall understand the limitations of production engineers if they express inability to overload the plant units; production engineers shall cooperate with maintenance crew by more carefully operating the plant or arranging to stop the plant for some urgent repairs; stores department shall make available spares in time for repairs. This can help achieve safety, pollution control, and product quality. An environment of TQM is thus built up and maintained in the organisation.

Training—Training is very important for employees to be highly productive. Management and seniors are responsible for implementing TQM within the organisation and their respective departments and hence should organise professional training programmes—which are the key by which the TQM succeeds in an organisation.

- Orientation of new employees and to explain how they can contribute to the goals, the processes, and operations they will have to carry out and building trust among all by honest, sincere, and ethical working,
- Refresher courses for old employees,
- Mock drills for all employees to train them for response to emergency,
- Familiarise working of different departments. Persons should be trained to be able to work at different workstations if required. This training also helps them understand difficulties faced by others; they may offer some useful suggestions for improvement since they will be looking from a different angle.

Employees should be trained in technical skills, analysis of problems, generating new methods and ideas, estimating benefit–cost ratios, improvement in performance, and requirement of costs and time to implement any suggestions/ideas and to work as a team by mutual trust.

Teamwork: Discussions and cooperation among operating and maintenance personnel (chemical, mechanical, electrical, material purchase and marketing) can bring about permanent improvements in processes and operations. The problems can get highlighted, and practical solutions can be found out. The organisations can build up teams of specialists and experienced personnel for continuously improving the working which contributes to the success of the TQM approach:

- Quality improvement teams—for addressing some recurring problems.
- Troubleshooting teams—to solve certain problems by identifying their causes and to minimise them.
- Working group teams—these could be small groups of skilled workers who share tasks and responsibilities. These can be quality circles also.

Recognition
It should be given for both suggestions and achievements for teams as well as individuals. Giving due credit to deserving person as early as possible for good efforts, ideas, and suggestions is important job of a senior. It improves productivity, quality, and the employee morale. Recognition can be as:

- Special increment in salary,
- Promotion to higher post,
- Letter of appreciation from senior management,
- Praised in front of colleagues, in-house journals, in general notice boards, also in front of very senior personnel, and in staff meeting,
- Special award in annual company functions.

2.15.3 *Suggestion Schemes*

Provide sealed boxes in every department where suggestion for safety, pollution control, Energy efficiency, better product quality, reducing breakdowns, reducing cost of production, increasing productivity of process equipments and machinery can be dropped in. These boxes shall be opened once every month and the best three suggestions can be implemented after making a benefit—cost analysis. Persons making these suggestions can be suitably rewarded.

Chapter 3
Procurement of Equipments

3.1 Introduction

Various types of equipments are required for setting up and running a chemical plant. It is necessary to prepare correct specifications for each one of them for the present needs and in future as per production capacity and product mix planned.

It is necessary to procure the right type of process units and reactors, machinery and auxiliary equipments for a safe, pollution-free, and efficiently running chemical industry. Care shall be taken to avoid incorrect specifications while floating enquiries as these can lead to waste of time for procurement (long delivery periods), the bought-out items not meeting some important required features, delays in erection or commissioning, spares not available or delayed in future. Difficulties may arise during operation and maintenance, and there could be need for occasionally inviting outside experts.

Purchase of high-capacity units in the beginning itself may not be the right choice as their operating costs per unit of output could be high if they are run at lower capacity initially. There may be additional costs for installation of such units.

3.2 Planning for Procurement

Following steps may be found useful by the readers.

- Finalise the product mix for the present and for the future (with or without expansion of the overall capacity of the industry). This shall be done for the manufacture of different products and their grades as well as rates of production of each type. *It is necessary to carry out a detailed market survey before this step.*
- Select the process know-how and technology.
- Make a list of all items of plant and machinery to be procured.

K.R. Golwalkar, *Production Management of Chemical Industries*,
DOI 10.1007/978-3-319-28253-4_3

- Prepare detailed specifications for all items by considering the product mix, scale of manufacture and technology to be used.
- The quality of raw materials (which are to be used), the local conditions of climate, available infrastructure, and quality of water at site shall be considered while procuring the equipments since there may arise need for some additional units. They may affect *the choice of materials of construction* also.

Finalise the process flow diagram and the piping and instrumentation diagrams. Carry out HAZID and HAZOP studies.

3.2.1 Common Equipments and Machineries

- Material handling for production unit—(for solid, liquid, and gases);
- Pretreatment facilities—crushing, grinding, calcinations, and filtration for the raw materials;
- Main process units/reactors and accessories—catalysts, tower internals;
- Pressure vessels and accessories (safety valves and rupture discs);
- Waste heat recovery boilers and economisers, feed water pumps, and safety valves;
- Water treatment plants with pumps, resin columns, membranes, and instrumentation;
- Refrigeration plants and cooling towers;
- Instrumentation;
- Electrical installations and motors, cables, control centres, and cogeneration units;
- Process and quality control laboratory facilities;
- Effluent treatment plants and disposal of wastes;
- Storages, handling of raw materials, and dispatch finished products;
- Fuel storage and handling;
- Infrastructure at site—maintenance facilities, transport fleet, and fire fighting;
- First aid and safety equipments.

3.3 General Guidelines

These are given below for floating enquiries for the various items. The enquiry documents should specify the duty conditions in detail for the present and in future. The capacity and intended use of the equipment, as well as requirements of any specific feature in the respective equipments shall also be mentioned in the enquiry documents.

(*Readers may refer to the book* ***Process Equipment Procurement in the Chemical and Related Industries*** *by this author for more technical guidelines.*)

- Numbers of each item/equipment required.
- The delivery period shall be quoted by vendor. The purchaser can inform if it is needed urgently.
- General arrangement (**GA**) drawing of the item while showing battery limits of scope of supply by vendor, and all input/output points.
- Codes and standards to be followed for fabrication.
- Overall size and weight (when empty)—**by vendor**.
  Plant engineers/designers shall estimate weight of the items with all fittings such as valves, connected pipes, and when full with process reactants/chemicals also. Vibratory load when running at full speed shall be taken into account for designing and fabricating. Corrosion allowance shall be added.
- Foundation drawing shall be given by vendor along with supply of foundation bolts.
- Materials of construction MOC of critical parts/wetted parts to be indicated. Vendor may be asked to suggest alternative material provided the required performance and life of equipment is guaranteed.
- The equipment shall be supplied as a fully assembled unit duly aligned with drive motor on a base plate to the site.
- As partially assembled (if internal fittings and inspections are to be done at site) and a separately supplied drive motor. Final assembly at site has to be done in the presence of the engineer deputed by vendor.
- Instrumentation shall be supplied by vendor (with the unit or separately).
- Erection and Commissioning shall be included in the scope of supply. Performance guarantee test shall be for 72–168 h of continuous run (3–7 days); or for producing five batches of the product as per specifications in case of batch process.
- **Battery limits of scope of supply** and exclusions from scope of supply to be made very clear by vendor. Maximum permissible speed, pressure, temperature, and feed rate shall be specified by vendors. (This can enable assessing capacity for future expansion also.)
- Requirement of utilities for erection, trial runs, and normal operation after commissioning and any special accessories required should be given by vendors.
- Test certificate of all bought-out components to be given by vendors, e.g. safety valves, motors, view glasses, heat exchangers tubes, and boiler tubes.
- **Quality Assurance Plan** to be given by vendors (Fabrication codes and standards, stress relieving, post-weld heat treatment in case of pressure vessels).
- Should have necessary licence/permission from statutory authorities for fabrication of pressure vessels.
- Vendors may be asked if they can offer an annual maintenance contract.
- Vendor shall inform exclusions from scope of supply.

3.3.1 Purchaser's Obligations

- The site facilities required for erection, commissioning, and satisfactory working of the equipments if requested by the vendor. This matter shall be very clear between the purchaser and the vendor.
- Attend stage inspection and final inspection *before despatch* if called by the vendor at his works.
- Inspection of all equipment after unloading at site in the presence of vendor's representative.
- Proper safe storage at site.
- Making civil foundations ready at site for placing the units.
- Making available unskilled labour for unloading, and internal shifting up to foundation and placing on it, assistance for handling heavy loads.
- Providing water and power for erection (these are generally demanded free of cost by the vendors).
- Thermal insulation and cladding (these may be included in vendor's scope).
- Obtaining necessary statutory clearances for erecting and commissioning.
- Making available unskilled manpower for performance guarantee test runs. Skilled operators and engineers *shall also be deputed* by purchaser during PG test runs since they should get trained in running the equipments.

3.4 Process Units/Machineries (Present Requirements)

These are equipments which are to be installed at present itself. They are run at reduced capacities currently but can be run at higher rates in future when production capacity is to be expanded.

Either no change or only a very little modification is required in these equipments in future.

- *Increase the present running hours of the treatment unit,*
- *Run the pump at higher delivery rate and open delivery valves further,*
- *Increase speed of agitator in the reactor,*
- *Add some more catalyst to the converter bed, and*
- *Add a few more plates to the plate heat exchanger.*

Example
The water treatment plant (WTP) is running 8 h per day at present when a sulphuric acid plant is producing 80 tonnes per day. In future, when the plant capacity is to be increased to 120 tonnes per day, the WTP will have to be run for at least 14–15 h per day (while assuming two hours for backwashing and regeneration). This is possible without modifying the WTP.

3.5 Units Required in Future

These are equipments required in future when the capacity of the plant is to be expanded or more products are to be manufactured.

- Some of the existing units can be used with modifications (e.g. increase height of absorber),
- Some of the existing units will have to be replaced by bigger units (e.g. bigger water pumps),
- New equipments which will have to be added.

Examples

1. The air drying tower (of sufficient internal diameter and lined with acid resistant bricks) in a sulphuric acid plant was having a packed height of 4000 mms of Intalox saddles and no more of them could be filled in. Replacement of the tower was considered to be very costly and would have needed a long time. It was therefore decided to increase the height of the tower by about 2000 mms to accommodate the additional Intalox saddles required to dry the additional

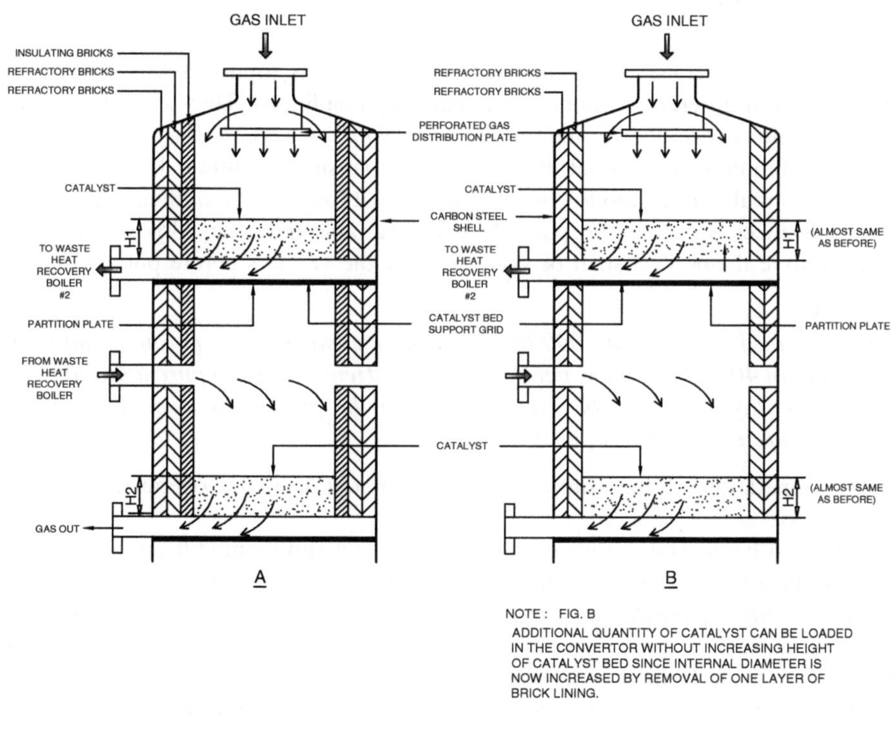

Fig. 3.1 Plant expansion with same converter

quantity of air required. Incoming flow rate of acid (for drying the air) was also increased slightly.

2. **Case study** Addition of some more catalyst to the converter of a sulphuric acid plant was done (without increasing the pressure drop) by removing one layer of brick-lining which was provided in the beginning itself. Please see Fig. 3.1.
3. A cooling water pump is running at about 30 % of its rated capacity at present when the plant is producing 80 tonnes per day. In future, when the plant capacity is to be increased to 120 tonnes per day, the pump will have to be run at about 45–50 % of its rated capacity. This is well within its capacity, and it can be easily run without any change.
   However, if the cooling water pumps is already running at 65–70 % of its capacity (when the plant is producing 80 tonnes per day), it will have to be run at more than 95 % of the rated capacity and this may not be possible. Replacement by a bigger capacity pump will be required in such a case.
4. Initially, only sulphuric acid is being produced. Provision of a new oleum tower with accessories such as circulation pump, coolers, and separate storage tanks will be required if it is planned to produce 25 % oleum also in future.

3.6 Selection of Equipments

It is advisable to consider various types of equipments (options) which could be available from different vendors for similar duty conditions. They should meet the duty conditions specified for the present *and as far as possible* for the future also. This will minimise additional expenses for expansion in future. However, their efficiency should not be too low when operating at lower range of their capacity. In many instances, this may be possible when variable-frequency drive motors are used to drive them. This shall be confirmed by the vendors before purchasing.

Example
A sulphur feeding pump of 1000 litres per hour capacity can operate a sulphuric acid plant of 40 to 100 tonnes per day capacity. Hence, a pump purchased for a 40 tonnes per day plant can be used even when the production capacity is to be increased up to 100 tonnes per day .

- Safe operation of the equipment even when it is to be operated at higher capacity.
- It should be possible to increase the capacity by simple modifications—adding more heat transfer plates to the unit (*Plate Heat Exchanger*), *change the design of the agitator, create more space inside the converter by removing a layer of refractory bricks to accommodate additional catalyst* (*please refer to drawing*).
- Shall be made from corrosion-resistant materials for longer life.
- Power consumption shall be low even when it is run at lower capacity (*bigger units consume too much power per unit output when run on low load*)

- Shall have high operating efficiency even under different load conditions. *Some tubes in a heat exchanger can be kept plugged when the plant is running at lower capacity. This is to ensure sufficient velocity of fluids in the tubes for better heat transfer. The plugs from the tubes can be removed when the plant is to be run at higher capacity* in future.
- Shall be easy to install in plant/shall not need heavy-duty erection equipments.
- Shall not occupy too much space which can obstruct other activities in the plant.
- Ease of operation and change of capacity during normal run of the plant. The drive motors shall have variable-frequency drives.
- Ease of maintenance.
- Ease of cleaning during annual shutdowns.
- Immediate availability of spares.
- Initial capital cost of the item, and the interest paid on higher capital till production capacity of the plant is expanded. This is to be weighed against lower funds required for the expansion in future.
- Operating cost of the item—this is to be specially looked into.

3.7 Some Miscellaneous Items for Making a Choice

Some miscellaneous items for making a choice are as follows.

3.7.1 Vertical Submerged Pumps

These pumps do not need priming when they are installed on tanks. Any leaks from glands fall back into the tank and there is no spillage outside.

Need gantry for maintenance work.

External Horizontal Pumps
Any leak from gland of externally installed pumps can be dangerous; needs acid-/alkali-resistant tile lining on foundation and surrounding area. Hence, they need mechanical seals for handling corrosive, inflammable liquids.

3.7.2 Refractory Bricks

Refractory bricks—the selection will depend on the service temperature in the furnace, the shape (special curved bricks are tailor made and cost more) and size; and intended use as lining of surfaces or as arches for roofs or manholes with 45 or 65 %

alumina. Iron oxide content shall not be more than 0.2 %. Vendor to submit test report for refractoriness under load (RUL). *Please see below for other lining materials.*

3.7.3 Caesium-Promoted Catalyst

Caesium-promoted catalyst for sulphuric acid plant has a lower ignition temperature and can result in higher overall conversion of $SO_2 \rightarrow SO_3$; but is more costly. It can reduce consumption of alkali in tail gas scrubber as compared to conventional potassium-promoted catalyst.

3.7.4 Heat Exchangers

(i) Acid coolers

(a) Shell and tubes of Stainless Steel—It can cool sulphuric acid, but cannot recover heat as hot water with a temperature approaching 3–4 °C of incoming hot acid. Heat transfer area cannot be increased for higher load.
(b) Hastelloy Plate Heat Exchanger—It has a very good corrosion resistance, needs much smaller space due to high coefficient of heat transfer, can get hot water up to 3–4 °C of incoming hot acid, and heat transfer area can be increased for higher load by adding some more plates.

(ii) Gas to gas shell and tube heat exchangers

These have disadvantages of reduced efficiency if orientations of the inlet and exit nozzles are not correct, have more pressure drop, and chances of idle pockets as compared to disc-and-doughnut-type units. However, cost of the latter can be more.

3.7.5 Pressurised Cooling Water Flow Arrangement

Ingress of cooling water can occur through a leak and can result in an accident.

Cooling Water Flow Under Siphon
Any leak will stop cooling water flow automatically as the siphon flow will break; hence, this is a safe system. **Please refer to drawings in Chapter on Safety Management. They are not repeated here**.

3.7.6 Tangential Spray Scrubbing Towers

These are used for pollution control in phosphatic fertiliser plants where deposits of silica can occur in scrubbing towers. These deposits need to be removed soon. The tower is mostly empty and has low pressure drop. The tangential sprays wash the tower walls continuously (to remove the deposits) and also keep the walls cool by flow of scrubbing liquor.

The liquor inlet pipes can be cleaned by inserting rods from outside.

Scrubbing Tower with Central Liquor Spray
Good scrubbing action when the spray covers entire cross section; but spray nozzles cannot be cleaned easily, unless they are taken out.

3.7.7 Teflon Gaskets

Teflon is a very inert material and resists most corrosive substances up to 230–240 °C. The gaskets do not stick to flanges and can be changed.

Example
Can be used in very corrosive service, e.g. handling vapours of pure SO_3 generated by boiling oleum in a liquid SO_3 production plant.

Compressed asbestos gaskets for corrosive service are not recommended for such service as they may leak.

3.7.8 Graphite Electrodes

Safer compared to water-cooled electrodes (where accident can occur due to ingress of water into furnace from a leak of water).

Example
These are installed inside a refractory-lined electric furnace to generate heat internally.

There is more efficient utilisation of electrical power (since no cooling water flow is required), but are more costly and delicate to handle. Need great care to store, install, and add to an old consumed piece (by screwing on a new piece). However, the installation of a new electrode requires more expertise. The repairing of damaged electrode is difficult.

An external water-cooled gland is sometimes provided to arrest leak of inside material.

3.7.9 Shaft Sealing Arrangements

- Conventional packed gland for rotating shafts.

It can get damaged due to friction; the leaking vapours present a fire and toxic hazard.

Special Shaft Sealing Arrangements
These are used (i) for pumps which handle toxic, inflammable, liquids (ii) for agitator shafts of reactors carrying out reactions inside while handling dangerous, toxic, inflammable materials. Some of these are as follows:

- Water-cooled glands
  Chances of fire/toxic leak are reduced but are not completely eliminated.
- Mechanical seals—costly, but are safer than above two. However, gas detector shall be provided in nearby area as a precautionary measure.

3.7.10 Centrifugal Blower

A centrifugal blower can deliver large volumes at lower discharge pressure, but cannot deliver steady constant volumes if there is high backpressure in the system. If it is necessary to have steady volumes of air (gas) delivered, this type of blower cannot generally do so, unless it is specially designed with multiple stages of impeller.

It can result in loss of production due to less air delivery in some plants.

Positive displacement (roots) blower can deliver steady constant volumes even if there is high pressure drop in the system. However, the drive motor will draw heavy current in such a situation; but there is less chances of process upsets or production loss.

Since the clearances between the impeller and volute are very small, there is considerable chance of mechanical erosion or jamming of the impeller if there are dust particles in the gases. Hence, protection devices like tripping of the drive motor are incorporated by the blower manufacturer in such cases.

3.7.11 Water-Tube and Smoke-Tube Boilers

Water-Tube Boiler:

- It is used generally for power generation and, hence, operated at higher pressure.
- Quicker start due to low water volume in the tubes.
- Since the operating pressure is high, the saturation temperature is also high. Due to this, the temperature of exit gases is also high (which is necessary to maintain

a reasonable temperature difference driving force for heat transfer from hot gases to boiling water in the boiler).
- Chances of tube leak are more.
- Cost of boiler shell cost is less, because shell is not subjected to high pressure.
- Ferrules are not required to be fitted in the tubes.
- Higher steam demand can be met in shorter time.

Smoke-Tube Boiler:

- These were used in locomotive steam engines.
- It takes longer time to start due to high water volume in the boiler.
- Cost of shell cost is more because shell is subjected to high pressure.
- Ferrules are fitted at the inlet of tube, and tube sheets are covered by refractory lining. These are required for protection of the tube inlets as well as the tube sheets.
- Higher steam demands cannot be met in shorter time.

Conventional economiser and Steaming Economiser, and various heating methods

—please refer to Utilities chapter.

3.7.12 Partition Rings

It can have more gas liquid contact area but can get choked due to suspended solids in liquid. **Intalox saddles** can provide good gas liquid contact, but offers a little more pressure drop.

3.7.13 Demisters

These are installed inside gas–liquid contact equipments for arresting escape of mist particles along with outgoing gases since they can corrode downstream process units (*when they are acidic*) *or can cause atmospheric pollution when the gases are released through a chimney.*

Pad-Type Demisters have low pressure drop but being light can get disturbed from position and hence need tying down on support frames. They have low separation efficiency for mist particles below five microns.

Candle-Type Demisters have more efficiency of removal of mist from gas streams (2.5 microns also possible), but have high pressure drop, and need more space to install.

3.7.14 Lower Portion of Towers as Circulation Tanks

Need less floor space, and there is no need of U-seal pipes for exit liquor from the gas absorber tower. Externally installed circulation pumps with mechanical seals are required.

Separate Circulation Tanks
Submerged pump can be provided; but separate U-seal piping is required for exit liquor to prevent gases coming out from the tank.

3.7.15 Protective Linings for Equipments

These are provided for process reactors, furnaces, gas absorbers, etc., for:

- Protection of these equipments against chemical corrosion and extreme heat (thermal protections to steel shells);
- Minimising heat loss from the equipments;
- Preventing contamination of products;
- For protection against chemical corrosion.

Materials used for lining:

- Insulating and firebricks.
- Castable refractory. The grain size shall generally not exceed 3–4 mms and should be able to withstand required service temperatures (1200 °C at least).
- Fibres of stainless steel 316 (of 2 mm diameter and 25–50 mm length) can be mixed with the castable refractory as advised by refractory engineer.
- Acid-/alkali-resistant tiles and bricks.
- Cladding of internal surface by 2.0–4.0 mm thick 316—stainless steel sheets.
- Fibreglass reinforced plastics.
- Neoprene rubber sheets of 3.0–4.0–5.0 mm thick as per specific need.
- PTFE (Teflon) sheets of 2.0–3.0–4.0 mm thick as per specific need.
- Glass lining of 2.0–4.0 mm thickness as per need.
- Lead lining/bonding.

Selection criteria for Lining Materials:

- Operating temperature inside the equipment.
- Material being handled, their compositions and corrosive properties, and their tendency to form sticky hard deposits on internal linings. Erosive properties of the reaction mass shall also be considered as the lining can get peeled off during operation.
- Whether any highly exothermic reactions are to be carried out inside and chance of thermal shocks to the lining due to local heating.

- In case of internal heating by oil firing/electrical heating, and when conducting slag may be formed, it is necessary to provide impervious and non-conducting lining. Mica sheets may be placed between the shell and the lining.
- Outside surface temperature of shell of the equipment shall not exceed 50–55 °C to minimise heat loss to surroundings by providing layer of insulating bricks also. *Brick-lining of equipments is to be done generally at site since the lining can get disturbed during transportation even over a short distance.*
  At very high tip speeds of blowers, the rubber/FRP lining can come off exposing the blower to corrosive gases. Please discuss with party who will carry out the lining job. This problem can arise in case if induced draught fans of pollution control units.

Test certificates for refractory bricks, cements, and castable lining material (their properties and compositions) shall be obtained. Some typical values are given below. Plant engineers may specify more stringent conditions to suit the operating requirements.

- Alumina content—higher Al_2O_3 % is able to withstand higher temperature.
- Porosity (firebricks <20–22 %, insulating bricks can be up to 60 % porous).
- Thermal conductivity K Cal/m °C.hr, and coefficient of thermal expansion.
- Refractoriness under load (RUL) in Kg/cm^2 at service temperature °C.
- Cold crushing strength in Kg/cm^2.
- Pyrometric cone equivalent (PCE) Values—*by vendor.*
- Iron content shall not exceed 0.15 % (higher iron content can lower RUL).
- Resistance to acid/alkali (% loss in weight in 24 h when exposed to operating conditions. This shall not exceed 1.0–1.5 %).

Typical Specifications for Bricks

Acid resistant bricks

- As per IS: 4860-1968 Class-I or equivalent international standards.
- Surfaces shall be smooth (without scratches) and edges shall not be broken or damaged. *Salt glazed surfaces are preferred.*
- Water absorption not more than 1.5–2.0 %.
- Compressive strength 700 kg/cm^2 (min).
- Resistance to the acid, i.e., loss in weight not to exceed 1.5 % in 24 h when in contact with the acid at extremes of operating conditions.

Insulating bricks

- IS: 2042-1972 type-2 bricks or equivalent international standards.
- Pyrometric cone equivalent: 27 (ASTM).
- Bulk density (max): 0.9 gm/cm^3.
- Apparent porosity: 60 %.
- Cold crushing strength shall not be less than 15 kg/cm^2.
- The brick shall be able to withstand 1250 °C temperatures on continuous duty.

High alumina bricks

- IS: 8-1983 type-2 or equivalent international standards.
- Shall contain not less than 45 % Al_2O_3 and the Fe_2O_3 content shall not exceed 0.3 %.
- Pyrometric cone equivalent (ASTM): 32.
- Refractoriness under load (RUL): 1400 °C.

Vents: It shall be provided on the equipments at suitable places for escape of moisture during curing/initial heating. This can prevent development of stress inside.

Curing of refractory lining: Slow uniform heating of the lining shall be done after the initial set has taken place by using a good amount of hot air. The temperature rise in first 24 h shall not exceed 150 °C and then 250 °C in the next 24 h.

Further heating to a higher temperature, maintaining it for 24 or more hours (*called soaking period*) and the gradual cooling shall be done as per advice from refractory engineer.

3.7.16 Rubber/Fibreglass Reinforced Linings

Important considerations (*conditions to which the lining will be exposed*):

- Analysis of fluids being handled, especially the presence of fluorine compounds.
- Operating temperature (normal) and maximum temperature likely to be reached.
- pH of fluid, acid, and alkali concentrations.
- Suspended solids/crystals and their abrasive nature (if any).
- Whether liquids being agitated will be violently in motion.
- Lining material shall be able to resist erosive and corrosive properties of material being handled. A small test piece may be prepared by the lining material and tested at actual operating conditions—by suspending in the liquid or in the gas duct.
- Electrical spark test. The lining must withstand at least 5000 V. Choose suitable thickness of lining accordingly (but it shall not be less than 5 mm for better life). Spark test indicates spots of weak lining.

3.7.17 Poly Tetra Flouro Ethylene (PTFE) Lining

This is a very inert material and hence not attacked by almost all chemicals in use. It can be used where product contamination must be avoided. PTFE sleeves will be found very useful for vessels, pipes, and valves in case of highly corrosive chemicals.

Limitations

It may not be used where the liquids may contain abrasive solids in suspension. It is not recommended above 250 °C and is quite costly. The heat transfer through PTFE lining is poor.

3.7.18 Glass Lining for Equipments

Advantages

- Used for manufacture of very pure chemicals, pharmaceutical industries, research laboratories, etc., since there are very less chances of product contamination.
- Process is visible and can be studied easily if glass equipments are used.

Disadvantages

- Poor mechanical and thermal shock resistance.
- Susceptible to mechanical vibrations, presence of abrasive or hard solids.
- Can get affected due to fluorides (especially in the presence of acids).

Special Precautions

- Special PTFE bellows and gaskets shall be used for connecting the pipes.
- Need anti-vibration supports to prevent damage to the glass-lined vessels and connecting glass pipes. Rubber pads can be used at support lugs. Borosilicate glass should also be treated with similar care as normal glassware.
- The glass-lined equipment shall be located in the premises away from vibrating machinery (such as compressors, crushers, and pulverisers), sources of heat (high-temperature furnaces). Avoid outdoor installation.
- Sufficient clearance shall be provided on all sides of GL equipments for movement of men, materials, and maintenance of equipments.
- Raw materials with abrasive particles can damage the lining.
- Any material (solid or liquid) shall be added very slowly to the process vessel to avoid hitting the glass-lined shaft, agitator, or vessel internals.

The following details are to be given to the vendor:

- Operating pressure, temperature, composition of the reaction mass (concentration of various components), viscosity, density, any heat evolved during the reaction
- Heating cooling and washing time cycles if batch process is followed.
- Any special type of gland required for the agitator shaft if dangerous materials are to be handled.
- Liquids feeding pipes to the glass-lined vessel shall be made of Teflon.
- Provision of a vent on the jacket for removal of vapours.

- Preferable to have an electronic speed governor for controlling speed of the agitator. A gear box may add any weight on the vessel, hence may not be used.
- Provision of Y-type valve (please refer to Chap. 6) for drain nozzle.
- Purchaser shall check the internals and surface preparation of the vessel before glass lining is done. Total thickness of the lining shall generally not be less than 2 mm anywhere.
- Vendor shall be asked to carry out that spark test at 5000 V at least for checking all the glass-lined portions of the equipments.
- The lined vessel shall be given heat treatment in a special furnace by slow heating for 24–48 h followed by slow cooling to relieve any residual stresses. The rate of heating/cooling and period of soaking shall be as advised by experts in glass lining jobs. Preserve the records of heat treatment activity.
- Vendor shall be informed of the operating conditions of the process—quantity of reaction mass, specific heat, initial and final temperature of reaction mass and the heating/cooling fluid in the jacket, in the glass-lined vessel as per process requirements. The vendor shall be asked to confirm the heat transfer by the fluid in the jacket. Decide the dimensions for heating/cooling jackets accordingly.
- Consult vendor for rate of heating and cooling to be followed: very rapid rates can damage the lining.
- High-speed agitators are not advisable as the glass lining can be damaged.
- Vendor shall be given the complete responsibility of loading at his works, transport, unloading at site, erection, and commissioning of the glass-lined vessel.
- Vendor shall give composition of the glass lining and guarantee for corrosion resistance to the reaction mass at operating conditions. A test coupon of the glass-lined piece may be obtained from the vendor and tested independently in purchaser's own laboratory.
- Purchaser must inform the vendor if any fluorine compounds are to be handled specially in acidic conditions as the glass lining may get damaged.

3.7.19 Lead Lining and Lead Bonding

Both need very good surface preparation and skilled workmanship especially when the process vessel has inlet and exit nozzles, charging manholes, etc. The lead must fully cover the nozzles and their welded joints with the process vessel.

Both offer good chemical resistance and can protect the process vessel. These vessels shall not be used for food industry as even minor traces of lead can be poisonous.

Weep holes (small holes) can be provided on the shell of the process vessel as passage for any leaking process liquid from inside (to minimise accumulation of the liquid between the vessel and the lead lining). The vendor shall be consulted if the vessel is to be used above 200 °C generally.

3.7.20 Valves, Flanges, Gaskets, Expansion Bellows, Bolts, and Nuts

These are required for fluid pipelines for transfer of fluids and their flow control; entry, exit, and drain nozzles of process and pressure vessels for a wide range of duty conditions.

While selecting these items, the operating pressures, temperatures, presence of corrosive materials, pH, shall be considered. Generally choose an item for a pressure rating more than or equal to but never less than the maximum likely pressure in the system. Please consult Chemical Engineers' Handbook. This is to minimise chances of leak.

Stress analysis of piping systems shall be carried out before selecting expansion bellows.

Chapter 4
Erection and Commissioning

4.1 New Process Plant

It is very important to take proper care during erection of process units, plant machinery, site fabrication of large-diameter storage tanks, absorption towers with internal acid-resistant linings and connections of pipelines; installation and calibration of instrumentation (e.g. calibrations of level indicators, thermocouples, flowmeters, reaction vessels), curing of the internal lining (of bricks or other material) while setting up a new process plant.

Mechanical trial runs shall be carried out as per instructions from manufacturers of the respective machinery and equipments before start of raw material feeding.

This can ensure smooth production runs, convenience during cleaning and carrying out maintenance; safe working while operating the plants and future expansion of plant capacity.

4.1.1 Future Expansion and Diversification

A well-designed layout shall have adequate space available for bringing in additional equipments for expansion of capacity or diversification to more products in future (with minimum disturbance to existing units and piping) during further erection work.

Word of caution

The team of senior engineers deputed by the production manager should monitor plant conditions and incidences like frequent blowing off the safety valves, heavy vibrations of stirred vessels, frequent tripping of motors (i.e. anything that appears abnormal) during commissioning/after a major change has been carried out in the plant. Immediate corrective actions can minimise delay in stabilising the production

K.R. Golwalkar, *Production Management of Chemical Industries*,
DOI 10.1007/978-3-319-28253-4_4

rate and achieving the product quality. *This is necessary when a new technology—* ***which is not yet proven****—is being implemented or a major change in the existing plant has been done (introduction of new equipment/new raw material), or another product is to be manufactured in the existing plant* when dangerous chemicals are to be handled.

- An unsafe situation can result if proper investigations are not done or suitable precautions are not taken immediately.
- Quality of product is not as per specifications even after a long time after starting the production runs of the plant (unable to sell even at low price) and storage space is getting filled up. *The plant cannot be said to have been commissioned successfully in such case.*
- Starting materials, other inputs, and raw materials are purchased as per old specifications being followed—and *purchase department is not aware of new specifications* which are given by the consultants/technology supplier in new contracts with management.
- There are frequent breakdowns in key machineries (pumps, blowers, conveyors) which are causing interruptions in plant operations.

4.2 Prerequisites for Erection

- Process flow diagram, P, and ID shall be finalised.
- Plant layout shall be frozen.
- Equipment designs with nozzle orientation shall be finalised.
- Routes of all piping decided.
- Locations of bends and supports for pipes almost decided (make 3D models).
- Routes of all electrical cables planned, and suitable foundations and supports for all high-voltage equipments made.
- Cables shall be away from pipes carrying inflammable, toxic, corrosive fluids. Any leaked out fluid shall not fall directly on cables.
- Storm water drains shall be made and kept clean always during erection activities. Equipments shall be inspected at vendor's works and brought to site—check that proper lifting lugs are provided on vessels—specially pressure vessels. These must be welded.
- The expected dates of delivery of major equipments to site shall be known or estimated after visiting vendor's works at reasonable intervals for monitoring the progress of fabrication.
- A list of fabrication jobs to be done at site shall be known such as storage tanks; large-sized process vessels like converters, absorption towers (which need internal linings); distillation columns (with internal fittings of trays).
- The requirement of resources and arrangements to be made at site (manpower, plate bending machines, special tools, heavy-duty cranes, welding sets) and the time required for erection of major process units shall be estimated. Erection of

certain units needs completion of some other jobs beforehand, e.g. the completion of civil foundations and receipt of internal fittings at site is necessary before erecting a converter unit.
- The project team shall work out the **critical path (CP)** for completing the erection work after the above estimates have been carefully considered.

4.2.1 Site Office Management: Important Activities

- Coordination with senior officers at head office on a daily basis to inform progress of site work, any extra funds or resources required, delayed activities, and corrective actions taken.
- Coordination with chief project engineer, consultants, and senior management at head office so that plant is erected properly.
- Liaison with statutory authorities for replying to any queries sent by them.
- The date of commissioning is generally known (or ordered by top management). Start working backwards from this date. This enables the project manager to focus on the activities which are lagging behind schedule. The **CP** shall be reviewed every day/week to find out activities which may delay the completion of project and plan for the next day/week.

4.2.2 Parallel Activities

- Levelling of land, making approach roads, Security arrangements, Stores rooms and office, weighbridges, arranging water and power for site fabrication, living quarters for workmen and their daily needs, first aid facility with an ambulance; making civil foundations and necessary support structures are the important activities which should be carried out (while equipments are being procured).
- Weights of all units and reactors shall be known so that civil foundations and necessary support structures shall be ready for placing equipments on them. The weights shall be calculated with all internal linings, fittings, loading of catalyst, tower packings, etc., as well as connected piping, external gearbox, drive motors, and vibrations during operations. Add wind loads as per height at which the units will be installed and operated.
- Scaffolding and temporary supports shall also be ready. Bring bending machines and welding sets to the site and arrange continuous power/DG sets for site fabrication.
- Power and water shall be available as per need—for construction, for testing vessels and pipeline, and for fire fighting.

4.2.3 Site Fabrication of Large Vessels (Storage Tanks, Absorption Towers, Converters)

- This can be started only after foundations are ready. Meanwhile, steel plates can be bent as per curvatures required and kept ready.
- Heavy-duty cranes, chain pulley blocks, and portable tripod stands shall be available.
- Refractory bricks and insulating bricks shall be stored in closed warehouse.
- Boxing up—fitting internal distribution trays, tower packings, catalysts, thermowells, and instrument cables.

4.3 Erection

Some likely mistakes during erection and their consequences are as follows:

- Foundations not fully cured before placing equipments on them.
- Equipments placed directly on the foundations and not on the I-beams which should be placed first on the foundation (the I-beams shall be below the equipments).
- Anticorrosive paint not applied on the bottom side of equipment before placing them on the I-beams/base frames.
- Nozzle orientations for incoming/outgoing streams not as per layout/as per piping diagram/designs.
- Expansion joints/bellows in ducts not provided.
- Inner shells of equipments not cleaned thoroughly; welding of nozzles incomplete, before starting refractory brickwork.
- Protective inner lining by refractory and acid-resistant materials and their curing must be done by properly trained and skilled workmen and duly supervised by competent engineers. Too fast heating for drying or not allowing sufficient setting time for the lining *before curing by acid* can weaken/damage the lining.
- Clean filtered water not used for making cement paste or for curing. (The total dissolved solids in water shall be less than 80 ppm.)
- Expansion joints in brick linings too thin/not provided at all.
- Insufficient time given for setting and curing of a layer of brick lining before laying the next layer—**moisture remains trapped inside and causes problems during operation due to cracks in lining or formation of incontrollable acid mist**.
- Process vessels erected too close to walls (not enough clearance from walls or permanent structures). In this case either the valves cannot be fitted, or the ducts cannot be welded properly; very less space becomes available for fitting field instruments, and it becomes difficult to operate, clean, and maintain the equipment.
- Escape routes and ladders too narrow, at incorrect places/difficult to reach.

- Process pipes in the way of ladders or escape routes; valve in process streams fitted at such locations where both operation and maintenance are difficult—can cause inefficient plant operation due to operator fatigue.
- Flanged joints not provided in long pipelines. Suitable drain points and valves not provided in liquid pipes; vent valves not provided in gas pipes **may become difficult to clean/replace pipelines**.
- Only a limited number of escape routes available—at least two independent routes should be available.
- Insufficient lighting on such routes.
- Railings not provided/too less in height.
- Tower internal distribution trays not levelled properly—can result in poor absorption or distillation performance.
- Internal spray nozzles not cleaned and checked before fitting (Fig. 4.1).

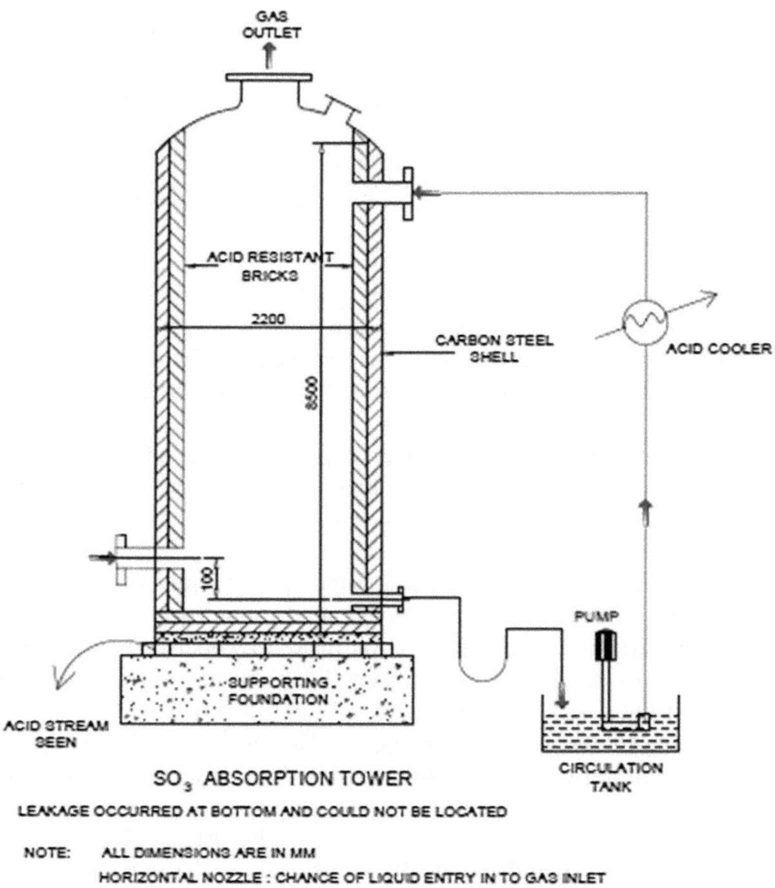

Fig. 4.1 Absorption tower (in a sulphuric acid plant) lined with acid-resistant bricks

- The angle of entry for the gas duct is faulty—liquid can enter the gas line if the tower exit is partially choked. The gas inlet line coming down from above shall be at 45° to the horizon.

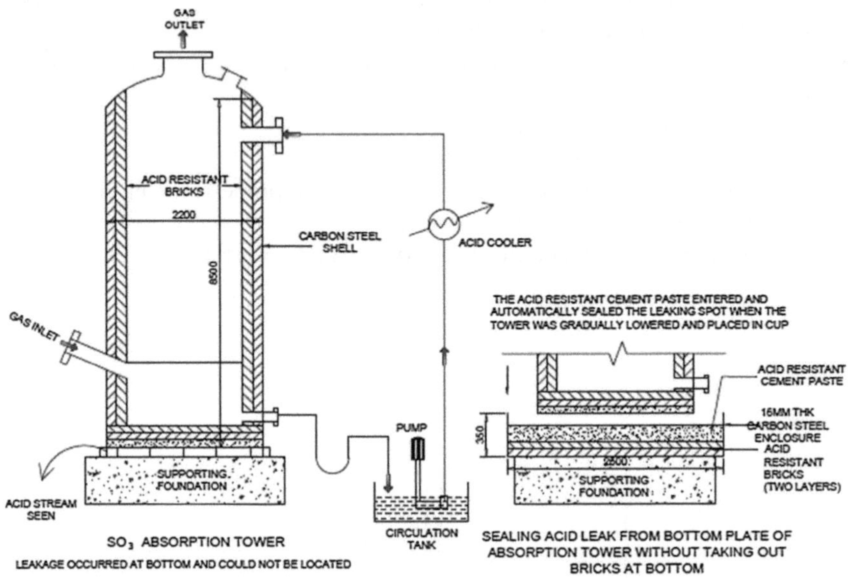

- Demister pads not fixed securely or gaps are left on sides—escape of droplets/mist with outgoing gas stream can cause damage to downstream units (interpass heat exchangers) or atmospheric pollution. Candle demisters should be fitted tightly with suitable gaskets on the tube sheets. *This is missed sometimes.*
- Welded manhole covers provided on process vessels instead of bolted covers—cutting and welding required every time for inspection or cleaning of the internals. This may be very difficult or unsafe if the vessels carry inflammable, toxic, corrosive fluids.
- Pumps are not properly wrapped and supported securely on wooden planks before storing if they have been received at site at an early date than required. Their nozzles are not kept closed by cover plates of wood/steel (dust can enter in the pump). It is advisable to put them back in the original packing after inspection if the pump is not to be erected immediately.
- Pumps of long submerged type not supported at intervals along their lengths and at the volute during storage (before erection). This may cause warping/sagging of the shaft.
- If chains are tied around the pump shaft or body instead of using the lifting hooks provided by manufacturer, the pump parts will get twisted or warped internally/life of bearings may get shortened.

- Separate supports not provided for connecting lines of pumps (suction and discharge); vent lines of safety valves (which can cause stresses in the pumps and safety valves).
- Sliding supports not provided for long horizontal furnaces at one end (they can slightly expand after commissioning of the plant).
- Instrumentation packing cases opened much before actual erection and then not closed properly causing ingress of moisture and dust, thus damaging the instruments.
- Instruments not stored safely away from dusty places.
- Storage room for electric motors and starters is not kept warm by low-power heaters or electrical lamps to prevent deposition of moisture (dew) in cold or humid weather. (The storeroom shall also be dust-proof.)
- Feed tank for metering pump placed at a height—*it can feed in liquid even when pump is stopped. This is not advisable. This tank shall be at a lower level.*
- Either strainers not provided at all or there are bigger holes in strainer screen in inlet line of metering pumps or instrumentation probe (concentration analyser).
- Sufficient straight lengths (as recommended by manufacturers) not provided before and after flowmeters like orifice meters/vortex meters.
- Bypass lines not available for rotameter, or for non-return valves in pipelines, making it difficult to take them out for cleaning or repairs.
- Insufficient thermal insulation on hot pipes can cause heat loss from process fluids and thus affect plant performance. It can also cause accidental burns to working personnel.
- Inlet and outlet ports of steam traps/non-return valves wrongly fitted in pipelines, *i.e. fitted in reverse direction* (*not as per marking*).
- Electric motor of wrong speed provided though it is of the same horsepower as specified.
- Frequency of power supply is 60 Hz, while motor is specified for 50 Hz.
- Guards not provided on gauge glasses or sight glasses of pressure vessels.
- Guards not provided on couplings/belt drives of rotating machines.
- Vent pipes of safety valves are not long enough—gases will get released in the working area or at low height, thus affecting personnel.
- Vent nozzles fixed at incorrect points on ducts or equipments (if this is not specified in drawings)—some trapped gases can remain inside.
- Drain nozzles fixed at incorrect points on ducts or equipments (if this is not specified in drawings)—some trapped liquids can remain inside.
- Flushing nozzles not provided on equipments or pipelines at appropriate places which will make washing and internal cleaning very difficult in future.
- Mistakes in understanding the colour codes for process piping.
- Cast iron pipes subjected to tensile forces during fixing flanged joints—spacer pieces of adequate thickness shall be provided if the length of pipe pieces is slightly short (say about 10–15 mm) instead of putting additional gaskets and tightening too much.
- Supports not provided for cast iron pipes at sufficiently close intervals—which can cause tensile stress in them and can reduce life of such pipes.

- Thin/wrongly designed flanges used on pipeline carrying dangerous, inflammable, toxic fluids or operated under higher pressure/temperature—**erection engineer must confirm that the flanges are as per recommendations given in Chemical Engineers' Handbook**.
- Material of construction of gaskets not compatible with fluids inside (strong acids) or operating conditions (high temperature or pressure).
- Internal fittings of converters, distillation columns, gas absorption towers, etc., should be done exactly as per instructions of designers and vendors of these equipments.
- Bolts of excessive length used for flanged joints—either the joints are not properly tightened initially or it becomes difficult to open them later on (for changing the gasket) because the threaded portion of bolts projecting outside the nuts can get rusted and jammed.

4.4 Precommissioning Checks of Infrastructure

- Capacity of raw water storage tank and how much of it can be actually available for use (check positions of overflow and outlet nozzles)
- Raw water quality, availability, supply rate, and supply timings
- Is there any other source of raw water?
- Water treatment plant and treated water storage tanks
- Main metering panel, maximum demand sanctioned, and capacitor bank installation
- Motor control centre, safe laying of all electrical cables and wiring, positions of starters and push buttons (these shall be at suitable places)
- Diesel generator sets and fuel tanks, batteries, and charging system
- Plant lighting—alternate lights to have supplies from separate feeders
- A functional maintenance repair facility with adequate stocks of essential spares
- Storage tanks for all products, fuels, and starting materials—calibrated and tested for leaks
- Fire fighting pumps, dedicated water tank, and other facilities
- Transport vehicles and weighbridge
- Process control laboratory with all necessary apparatus and analytical chemicals
- Equalisation tank and effluent treatment plant, stock of flocculants, lime, and alkali
- Air pollution control facilities and tall chimney (with all necessary fittings like aviation warning lights and sampling points) as required by statutory authorities
- Testing and calibration unit for instruments
- Refrigeration plants, cooling tower, and makeup water supply
- Packaged boiler with accessories, fuel supply tank, and chimney
- Lining of plant floor area by acid/alkali-resistant tiles with slope towards storm water drain and water taps for washing any spillage (provision of sand and ash buckets shall also be made if required)

- Roads for internal movement of personnel and machines and materials
- Warehouse for raw materials, forklift trucks, conveyors, and material handling facilities
- Dedicated water supply tanks for fire fighting, eyewash fountains, and safety showers
- Anti-skid gratings at appropriate places
- First aid, ambulance, and emergency conveyance facilities
- Workers living in quarters, restrooms, canteen.

4.5 Items to Be Examined Before Commissioning of Equipments

Process equipment or machinery not properly commissioned can result in unsatisfactory production rate, product quality, serious breakdowns, or serious accidents.

4.5.1 General Checkpoints Common for All Equipments (Prior to Their Commissioning)

- Civil foundation must be cured and inspected by civil engineer.
- Structural supports must be securely grouted, braced, and ready to take up loads of process vessels with internal fittings and process materials, lining materials, any hold up inside due to clogged exit nozzles, etc.
- Welding of all supports and bars shall be complete before placing any load.
- Foundation bolts must be of proper size, grouted securely, and provided with lock nuts, so they will not become loose due to vibrations when the machinery is running.
- Tag number of equipment actually positioned on a foundation shall tally with data sheets (for similar looking equipments) so that **wrong** equipment is not taken in use.

All pressure vessels, boilers, economisers, air receivers, fuel storages, etc., must have statutory approvals.

- Test certificates for all bought-out components shall be obtained from vendors.
- **All belt drives must have safety guards**.
- Speed, HP, frequency, supply voltage, supply phases, and frame size of all the motors which are fitted *confirm that they are as per specified on the machinery name plate.*
- Overall dimension as per final drawing.
- Visual external inspection of all fittings and mountings.
- Nozzle orientations for all incoming and outgoing fluids.

- Provision for inserting or connecting probes for the measurement of temperature, pressure, and flow of fluids.
- Sampling points with valves.
- Inspection windows, sight glasses and light glasses, and level indicators with guards.
- Settings of safety valves (shall have independent supports for connecting pipes to prevent warping of safety valves) and rupture discs.
- Drain points and vent nozzles with valves.
- Arrangement to flush out all residual material from inside.
- Test certificates for pressure parts, MOC, internal lining materials, etc.

4.5.2 *Pumps*

Various types of pumps are to be used in the plant. Confirm that the correct pump has been provided for the required service conditions. The selection of pump, material of construction of wetted parts depending on the liquid handled, discharge capacity and head, oil seal, water seal, and gland packing should be checked. Direction of rotation of pump and motor must match, and their alignment with each other must be perfect.

4.5.3 *Blowers*

Direction of rotation of blower and motor must match, and their alignment with each other must be perfect. The orientation of discharge nozzle shall be as per purchase order. It should not need many bends or changes in ducts on discharge side. Large-diameter ducts must be independently supported without putting load on the blower. The suction and discharge-side filter-cum-silencers shall be clean and also be independently supported. Motor must have protective cover. Orientation and position of junction box of the motors must be easily approachable for repairs.

4.5.4 *Electrical Items*

Motors, switch gear, transformers, cables, etc., shall have approval from competent electrical inspection authority.

All such equipments should have stood the prescribed necessary tests like insulation, winding integrity, and resistance with shells before switching on power supply to them.

4.5.5 *Heat Exchangers*

One should check that (i) all foundation bolts are properly fixed; (ii) provide ferrules of stainless steel or ceramic to ensure that the fluids entering the tubes of heat exchangers will not affect the welding with tube sheets (iii) protective lining on the tube sheet itself (iv) check nozzle orientations for entry and exit of process materials, liquids, gases and (v) there will be ease of operation/maintenance as well as for fitting of connecting ducts and piping.

Plate Heat Exchangers—Gaskets in paths of liquids should sustain the high temperature, pressure, and corrosive as well as erosive properties of the fluids. ***Confirm that correct gaskets are fitted***. The support frame shall be well fixed on the base frames, and working space shall be available to remove plates for cleaning during plant overhaul. Strainers may be required for the entering fluids if they contain particulate matter as these can clog the narrow passages in the PHE.

4.5.6 *Boilers*

These are the steam generation equipment. The boilers should have the various accessories like BFW pumps, safety valves, main steam, vent and blowdown valves, fuel feeding, level indicators and control arrangement, combustion air, and induced draught fans in perfectly working order. Ensure that instruments for indicating the parameters like furnace temperature, pressure, and water level are installed properly. The coal crushers, coal screening arrangement, belt conveyer system, and coal feeders should be checked for their proper functioning. All necessary statutory permissions must be obtained before commissioning.

4.5.6.1 Steam-Heated Units

Pressure tests of pipelines, jackets, steam coils, and arrangement to isolate and take out coils for maintenance are necessary. Steam traps shall be selected for the removal of condensate as soon as formed. *There shall be no cooling of condensate in the jacket or coils* (*unless system is specially designed to recover heat from hot condensate also*).

4.6 Pressure Vessels

- Confirm tag number and duty conditions;
- Check all test certificates for material of construction used for fabrication of the vessel and hydraulic pressure tests carried out earlier;

- Records of radiography test and post-weld heat treatment done;
- Confirm the settings of safety valves and fitting of rupture discs;
- Smooth operation of all valves—including vent and drain valves;
- Calibration of pressure and temperature gauges;
- Cooling water/chilled brine flow rate;
- Mechanical seal for agitator shaft;
- The connecting nozzles with safety valves and Rd shall be clean, and all exit and vent pipes shall be supported independently;
- All flanges, gaskets, sight glasses, and light glasses shall be tested and certified for operating conditions;
- Support legs and foundation bolts shall be as per maximum load expected with all external fittings, gearbox, agitator, raw material loaded inside, etc.

4.7 Distillation Units

Test for any leaks from shell and heating coils/tubes. Check that all the bubble cap trays are installed properly. All the instruments for flow measurement and control, temperature, pressure, and level indication on trays including the composition control unit (reflux ratio controller) are to be calibrated, and their probes shall be installed correctly.

4.7.1 Condensers

- Orientations of vapour inlet and condensate outlet nozzles.
- Orientations of cooling medium inlet and outlet nozzles.
- Check flow rates of cooling water through each of the condensers by observing the exit stream from each condenser separately (***the exit nozzles shall be disconnected from the common header for this purpose***—in case they are connected to a common header).
- Vent valves for release of non-condensable gases. The vent pipes shall be connected to a suitable exhaust system or to a safety vent.
- Sampling points for vapours and condensates.

4.8 Refrigeration Systems

Circulation of chilled water/brine and cooling water in the system, all safety valves and interlocks (as per technical manuals), test certificates for all pressure parts, leak test of refrigerant for the entire system, and capacity controllers for the plant shall be checked before commissioning. Cold insulation shall be made by using only those materials approved by statutory factory inspectors.

4.8.1 Cooling Towers

Check the basin for any leak. Check the water-level controller in the make-up water line for the cooling tower. Check all the fans, gearboxes, sealing arrangement, lubrication, etc. Check the motor loading (FRP blades are more efficient than metal blades).

4.9 Reactor with Agitators

The MOC of the agitator and shaft and gland packing should withstand the vapours and the reaction mixture itself. Examine the free rotation of agitator, proper installation of gearbox, belt drive (if provided), the speed reduction ratio, and the motor HP as well as speed of revolutions per minute (RPM). All these shall be as per original design—too high speed of agitator may overload the motor, may shear off the shaft, or may spoil the product quality. There should be sufficient space around the reactors, and there should be proper provision in case of maintenance. Use removable baffles if possible.

Examine the support legs and their reinforcements which must be able to withstand the weight of the units with all internal linings, fittings, loading of reactants, catalyst, etc., as well as connected piping, external gearbox, drive motors, and the vibrations likely to occur when the agitator rotates at full speed..

Check the installation of probes and sampling points for temperatures, pressures, and level indication.

4.10 Water Treatment Plant

Ensure the quality of raw water from the laboratory. Check the bed height of sand filters, fitting of RO membranes, exchange capacity of resin in the exchanger columns, working of sodium sulphite dosing pump, high-pressure raw water pump, air blower, etc., as well as working of the TDS and conductivity meters for ensuring the quality of treated water. The volumetric capacity of all tanks shall be confirmed to be adequate to supply treated water to process plant, boiler feed, etc. Drain line shall be available to purge out rejected concentrate. Pressure gauges shall be provided for all pumps: inlet and outlets of all units. Flow rates and level indicators for all treated water tanks shall be calibrated.

4.11 Effluent Treatment Plants

Volume of equalisation tank shall be sufficient to accommodate the effluent generated in 24 h. Ensure that chemical dosing and mixing system is working. Sufficient stock of lime and alum stock shall be available. Check flow through primary and secondary clarifiers. Pressure gauges shall be provided for all pumps: inlet and outlets of all units.

4.11.1 Aeration System

Check the depth of aeration nozzles below maximum water level. Run the blower and measure the discharge pressure as well as load on the drive motor at maximum air delivery.

4.12 Melters for Solids

Grids at loading points shall not have too big openings because it is difficult to melt big lumps (which may pass through the grid). Steam coils and jackets shall be tested at a pressure at least 1.5 times the maximum operating pressure. Pressure-reducing valves and safety valves shall be provided for steam inlet lines and set accordingly. Provide steam traps on individual coils and piping to collect condensate.

All chambers/sections of the melter shall be calibrated by measuring internal dimensions after internal lining job is done. The positions and dimensions of overflow and underflow passages shall also be checked.

4.13 Calciners

Check arrangements for feeding of wet solids, oil/gas firing and heating; working of ID fans; removal, cooling, and packing of calcined product as well as the height and supports of exhaust chimney.

4.14 Pipelines for Liquids and Ducts for Gases

Stress analysis shall be carried out for pipelines carrying high-temperature fluids. Ensure that the pipe supports and bellows are properly designed and installed in the pipelines at appropriate places to take care of thermal shocks, and gasket material and nut bolts should be made up of corrosion-resistant material. Roller supports may be provided for horizontal movements of pipelines. Consult piping engineer for correct advice in such matters. *Non-return valves shall be installed with bypass line and with isolating valves to enable taking them out or for in situ maintenance.*

4.15 Dry Compressed Air for Instrumentation

Check operation of air-drying plant (including proper switching on and off the compressor motor as per draw of air from outlet) and the complete pipelines for any leak. *The pipelines should be strong enough to withstand high pressure.* Check all air filters.

4.16 Precommissioning Checklist for Items: Case Study of a Sulphuric Acid Plant

1. Overall dimension of all equipments/units shall be checked, e.g. sulphur pit, converters, hot gas filter (HGF), acid towers;
2. Orientation of all nozzles for incoming and outgoing gas/liquid, e.g. gas inlet and outlet nozzles on HGF, heat exchanger (air preheater);
3. Civil foundations of all units;
4. Working space around each unit, operational and maintenance ease for all valves, instruments, etc.;
5. Foundation bolts of all units/motors, blower and whether they are properly grouted and secured in place;
6. Sulphur Pit: Calibration of each section of the pit; position of each overflow and underflow; covers and grids on the pit; hydraulic test for each steam coil;
7. **Sulphur Furnace and Vestibule**: Fixed and roller supports, inner air distribution, peepholes and manhole position, location of thermocouples, hydraulic tests for sulphur feed line, and sulphur spray gun for the furnace;
8. **Waste Heat Boilers**, **Steam Drum**, **and Superheater**: safety valves, level indicators, vents, fittings, non-return valves, water gauge glass (should be visible from control room) position of blowdown vessel;
9. **Boiler Feed Water System**—Check the demineralised water plant units, water piping, operation of water pumps, condensate recycle lines, deaerator tank, drain lines, calibration of Boiler Feed Water tank, operation of level indicator, external insulation, location of temperature indicators, etc. Get examined the hydraulic test certificates for all pressure parts and subsequently obtain permission from Statutory Boiler Inspection Authority to commission the boiler. **Note**: BFW pumps shall be run one by one to feed water into the boiler. Backpressure shall be created by throttling overflow from WHB. Water flow

shall be collected in a 200-l drum, and the discharge rate is measured against maximum operating pressure while checking current in each phase of BFW pump motor. Plant heating *shall not be started* till the working of the BFW pumps is found satisfactory.
10. **HGF**: Check position of inlet gas distributor and thickness of each layer of brick pieces. There shall be no pieces of insulating bricks. Clean bottom of HGF by sweeping;
11. **Converter**: Thickness of each pass of catalyst and check whether it is adequately supported by and covered by brick pieces/quartz pieces. Check position of inlet gas distributor, each thermocouple, and orientation of all nozzles (gas inlet/outlet);
12. Working of air blower(s) by blowing air through all units;
13. Working of acid pumps—(this will need an externally procured supply of acid);
14. Cooling water pumps for acid coolers.

4.17 Preliminary Checking of Infrastructure: Case Study of a Sulphuric Acid Plant

1. Internal roads and storm water drains;
2. Raw water storage and analysis of raw water;
3. Water Treatment Plant: visual inspection of units [DM and water softening];
4. Boiler feed water tank, boiler blowdown line and vessel; dilution water tank (overhead); chemical dosing tanks; makeup water line to acid cooler basin;
5. Fuel oil tank [main storage and day tank];
6. Oil Firing System: oil burner, oil heater and thermostat, and primary, secondary, and tertiary air lines;
7. Sulphur Storage Yard: approach road, fire fighting facilities, payloader;
8. Weighbridge (main);
9. Acid Storage Tanks: Dyke walls, AR tile lining around storage tanks, and breathers for each tank;
10. Acid dispatch arrangement (metering tanks, loading pumps, sampling arrangement, AR brick lining on working area);
11. Caustic soda solution storage (main tank and day tanks);
12. **ETP**: alum and lime dosing tanks, agitator, air compressor, pH controller;
13. Arrangement to receive materials like fresh 98 % Sulphuric acid, caustic lye, alum, fuel oil, raw water in the plant at appropriate points so that plant can be smoothly started;
14. **Emergency Power**: Diesel generator set, diesel storage tank, start-up battery, power change over switch;
15. **Chimney**: Total height of discharge point above-ground level, inner lining, aviation warning lamp, lightening arrester, sampling, and drain points;

16. **Stores**: Essential stores;
17. **Main incoming Power Supply**: Metering panel, HT transformer, barbed wire enclosure;
18. **Basin below Acid Cooler pipes**: total capacity, makeup water and drain lines, location of water pumps, etc.

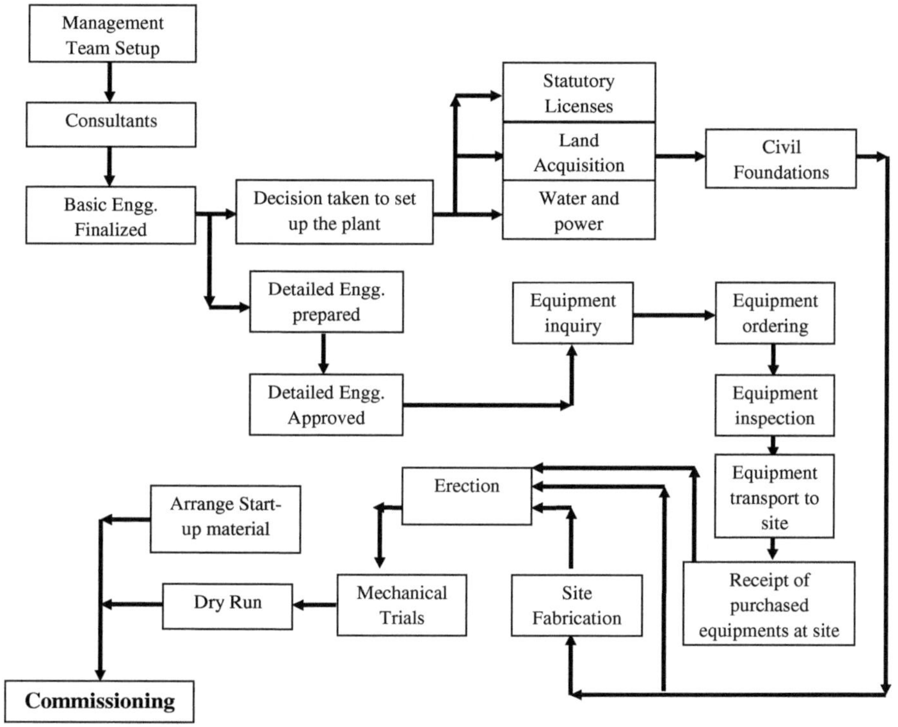

Progression from basic engineering to commissioning

Chapter 5
Procurement of Raw Materials and Other Items

It is one of the most important managerial responsibilities to ensure the availability of all necessary inputs for production. It is very important to procure them as per required specifications—especially the raw materials in the right quantities for smooth efficient production runs.

Some materials required for the production of chemicals are the finished products of other industries. These could be inorganic chemicals, organic chemicals, and solvents and are generally available as per specifications. They can be fed into the process units after confirming their quality by lab tests; they are sulphuric acid, caustic soda, ethyl alcohol, aniline, sodium sulphite, etc. These could be stabilisers and anti-caking agents also.

Any deviations from the specifications *beyond permissible limits* can disturb plant operations, cause breakdowns of pumps and choke filters frequently, increase utility consumptions (more steam will be required for drying wet raw materials), or can result in environmental pollution.

5.1 Selection Criteria

These depend on the intended use and hence the specifications for raw materials and other items shall be fixed for obtaining satisfactory marketable final product, without causing environmental pollution, or without being detrimental to plant equipments such as pumps, catalysts, heat exchanger tubes, etc. (due to the presence of undesirable impurities).

However, certain specifications like moisture percentage can be diluted, i.e. materials having higher moisture may be accepted when they are to be used in solution form or will get dried during exothermic reactions/heating during process.

The specifications and tolerance limits must be well defined with respect to the properties of the raw materials such as:

K.R. Golwalkar, *Production Management of Chemical Industries*,
DOI 10.1007/978-3-319-28253-4_5

- Minimum purity—*which is a must*—required (for sulphur granules for sulphuric acid plant) and maximum content of ash, organic matter which shall never exceed (may be somehow tolerated if the sulphur is not available from other sources).
- Minimum alumina content required and maximum moisture content allowed (in bauxite for alum plant).
- Maximum iron content allowed in acid used for paper maker's alum.
- Maximum chloride content (which can corrode many equipments, including SS-316), maximum calcium salts (which increase concentration of insoluble), etc. shall be clearly specified. *It is important to clearly mention that the presence of such impurities is not allowed as they can contaminate the product or corrode some of the equipments.*
- Maximum lump size allowed (big lumps of rock phosphate will take more time/consume more power for crushing and grinding in fertiliser plant).
- Minimum pour points and maximum viscosity (for ease of storage and handling of fuel oils) shall not be beyond the specified values. This is required for smooth handling of the oils in the plant.
- Likewise, the necessary compositions and tolerance limits for compositions of items like lubricants, effluent treatment chemicals, refractory bricks, and chemicals shall be worked out to ensure proper working of the process units and machineries.
- Chemical stability to mechanical shock, rain, and exposure to sun shall also be known. Appropriate precautions are required for safe handling and storage.

5.2 Rate of Consumptions and Inventory

A certain minimum stock level of all the necessary inputs (*maximum production rate per day × quantities required per unit of product × number of days required to procure* ***plus some extra stock as safety margin***) shall be always maintained so that there will be no shortage of such items. However, certain items can be very costly, and hence they shall be bought after a careful study of the data for the total production, actual sales, and actual consumption for the last six months to a year. A reasonably accurate forecast of the demand for the products will be found very useful for the procurement of the right quantities of such costly items.

The quantity of raw materials shall be thus sufficient to run the plant as per need (to fulfil commitments made for supply of products, to maintain smooth operations for a continuous plant, and to ensure safe and pollution free operations).

Special additives: *Anti-caking agents are added to hygroscopic products* (*if they tend to form cakes due to absorption of moisture*) *for ease of handling by screw conveyors, rotary valves.* The anti-caking agents maintain the free-flowing properties. One should ensure that these agents do not interfere with the reactions or cause corrosion of equipments. These items are required in small amounts but are important for the finished products. Hence, they also need to be procured in sufficient quantities to maintain production level.

Reorder point for each important material shall be checked and revised if necessary.

However, excessive inventory of inflammable, dangerous, toxic items shall be minimised.

A track shall be kept on availability of cheaper materials in certain seasons; from some particular source; lower transport charges and other such favourable factors. The aim shall be to minimise the landed cost at the site.

Procurement schedules for the materials shall be worked out in consultation with senior engineers so that the production and maintenance programmes planned by them can be carried out smoothly.

5.3 Acceptance of Items Which Do Not Meet Specifications

The production and maintenance engineers shall be consulted to see if some of the specifications can be relaxed (without seriously affecting the production rate, product quality equipment life, etc.) in case delivery of the material *as per required specification* is delayed.

Some typical cases are:

- Whether material with a higher moisture content can be used by providing a dryer or will it get dried during the process itself **OR** a somewhat higher steam consumption for melting the material can be accepted, e.g. a moisture content of 1 % can increase steam consumption of sulphur melters by about 19–20 %.
- Whether material with higher dust content can be used by washing or filtration, e.g. coal with more dust is washed before feeding to furnace and molten sulphur is filtered before use.
- Whether lumps of big size can be accepted by preliminary screening before crushing and grinding (prior to feeding in reactors).

However, certain impurities are very difficult to remove prior to feeding the raw material to process plant and they can affect the product quality, e.g. high iron content cannot be accepted in battery grade sulphuric acid and high chloride content in NaOH is not acceptable for rayon manufacture.

- Some of the impurities can spoil the product and waste the raw materials.
- Life and activity of catalyst: high dust content can mask the active surfaces of catalyst which can reduce reaction rates or conversion efficiency (conversion of input reactants to products). The escape of reactants from stack gases can result in pollution.

- Deposits of dust/ash on heat transfer surfaces (Condenser tubes, heat exchanger tubes) can adversely affect plant efficiency.
- Dust can also choke filter screens, catalyst bed/tubes, connections to probes for pressure measurement.
- Erosive dust particles can cause wear and tear of pumps, blowers, centrifuge and grinders by deposition of fine dust from the cheap dusty raw materials.
- There is always a possibility of corrosion of electrical bus bars, cable connection or motor cooling fans due to corrosive wet dust.
- Wear and tear of pump shafts can occur due to deposition of erosive dust in gland packing or due to ingress in pipes.
- Wear and tear of protective linings (glass/fibre glass) can occur due to the presence of sharp metallic pieces, grit and stones.
- The glass lining can also get adversely affected due to the presence of fluorides in the reactants if the reactor is operated in acidic conditions.

However, if the off-spec raw material can be pre-treated by cleaning, screening, filtering and drying at a reasonable cost before feeding to reactors without adverse effect on the production rate, product quality or equipment life, then such materials may be procured if they are considerably cheaper than the better quality but very costly and scarce raw materials.

The cost of pre-treatment shall include the capital invested in the equipments, cost of operation and maintenance also.

5.4 Storage of Raw Materials at Production Unit

- The materials shall be procured in such a way that there is no interruption in the production activities, i.e. adequate stocks shall be always available without excessive storage.
- Raw material bags can be stacked, say, ten bags in a row and five bags on top of each other. This makes it 50 bags per stack and it will be easy to keep track of arrival of fresh material as well as issue of raw material bags to process.
- The floor shall be water proof cement concrete/with acid resistant tiles and anti-skid grating. There shall be slope towards storm water drains. Sufficient ventilation, lighting (use translucent roof sheets) and exhaust system shall be available.
- Proper segregation of bags having solids and drums with liquids shall be done.
- Lubricants, fuels and inflammable materials shall be stored in separate area.
- The quantities ordered and their schedules of arrival at site shall be well planned to avoid problems such as filling volatile liquids up to brim in storage tanks (which can cause escape of vapours from the vents or create a fire hazard).
- It is better to keep a storage tank empty always so that the fluid from a leaking tank or container can be immediately transferred—thus preventing an accident.

5.5 Types of Packing for Procurement

- Purchase in small cans or carboys instead of full tanker loads if only small amounts are required. These can then be used directly in process units.
- Certain materials like sodium sulphate and lime powder are packed in small bags and conveyed by truckloads/railway. They can be unloaded manually or carried in hand carts at site if small amounts are required for use.
- If the materials are packed in big bags, they need unloading machines, portable hoists at site for taking out from trucks or railway wagons.

5.5.1 Purchasing by Tankers

- Instructions are to be given that a tanker should used be only for a particular chemical. If it is to be used to transport some other chemical also, then the tanker shall be thoroughly cleaned before every consignment. The material of construction and wetted parts of tanker and outlet valves should be compatible with the chemicals filled in, e.g. rubber-lined carbon steel for hydrochloric acid and stainless steel for nitric acid.
- The tankers shall have mountings (such as safety vents, pressure gauges, protection against lightening and a compressor to empty out the contents) and any other devices as per statutory requirements and must have licence to carry the cargo and permission to travel by a particular road.
- There is a possibility of development of high pressure inside when the tanker carrying volatile solvents may be waiting in the sun for long time. Hence, provision shall be available for external insulation, or cooling arrangement for the tanker. Such tankers shall be treated like a pressure vessel and all necessary precautions taken.
- The tanker driver must have safety manuals with necessary instructions for precautions to be taken in case of any leak. The telephone number and address of supplier shall also be painted.

5.5.2 Solids/Powders of Special Materials

Small amounts of these are generally available in mild steel drums, big plastic lined bags, small drums or packing made from thick paper sheets. Instructions from the manufacturer for method of unloading, storing (avoid direct sunlight, high humidity areas, etc.) and precautions for opening the packing must be followed always.

Avoid use of sharp knives and crow bars for opening the drums unless permitted by the supplier. There are chances of fire or explosions if metallic chisels and steel hammers are used. Use only non-sparking type devices (wooden mallets) with proper earth connection to each container (especially for Aluminium Powder).

5.6 Catalysts

Catalysts increase the rate of desired reaction (increase yield of desired products) and may allow the plants to be operated at lower temperatures, pressures, etc.

This can reduce the operation and maintenance cost of the plant.

5.6.1 Considerations for Purchase of Catalysts from Manufacturer

- The catalyst shall promote the desired reactions at **operation of the process at appreciably less** temperatures, pressures, etc. *Operation at lower temperature can increase overall equilibrium conversion in case of reversible exothermic reaction* (oxidation of SO_2 to SO_3 in the presence of caesium-promoted vanadium pentoxide catalyst used in sulphuric acid plants).
- It should be easily separable from the products and the unconverted reactants.
- **Reactor Operation with Fixed Bed of Catalyst**: Shape and dimensions of the catalyst particles should offer minimum resistance to flow of gaseous reactants even at a slight overload operation. This will reduce the power required to run the air blower for the plant (catalyst beds with hollow ring type catalyst have much less pressure drop as compared to the catalyst bed with solid cylindrical pellets used earlier in sulphuric acid plants).

5.6.2 Fluidised Bed of Catalyst

The process is sometimes operated with fluidised catalyst bed to achieve isothermal conditions. However, more energy is required for increased gas flow velocity and loss due to attrition are also more. The crumbled catalyst particles can get blown off and deposit as a fine powder on heat transfer surfaces of the downstream equipments. Cyclone separator is installed to remove the catalyst particles from gas stream

Attrition loss: Catalyst particles (pellets/spheres) shall be hard enough to withstand the loss during transport, loading in converter, screening during annual shutdowns and reloading in the converter. They shall not erode due to flow of gases during normal operations. A certain crushing strength shall be a specified bed or deposited in downstream equipments.

5.6.3 Process Conditions Leading to Loss of Activity

- Overload operation (high temperatures, higher concentrations of reactants) of the plant for meeting increased demand for the product due to which *sintering and consequent loss of activity of catalyst may take place.*
- High moisture in feed gases.
- Avoid the presence of certain impurities which can deactivate the catalyst. Consult catalyst manufacturer which impurities should never be allowed to enter feed streams of reactants since they may "poison" the catalyst.
- Moisture in feed stream, the presence of gases like Cl_2, CO_2 or thermal shocks due to frequent starting and stopping of plant can be detrimental to life of catalysts.

5.6.4 Initial Cost and Royalty

Compare the offers by various manufacturers regarding the initial cost of the catalyst as delivered at site, and the royalty payable for a patented catalyst.

Enter into a contract for technical services and guidance for optimum loading, commissioning, operation, screening, testing of samples and advice for replacement of the catalyst from the catalyst manufacturer.

Optimum loading of catalyst depends on converter geometry, internal dimensions, properties of reactants fed in and their concentration, flow rates of process streams and degree of conversion required for a given production rate of the product.

5.6.5 Handling and Storage

Obtain advice for these matters so that the catalysts do not lose their activity due to faulty storage (certain hygroscopic catalyst particles may get agglomerated if exposed to humid atmosphere and the active surface area gets reduced). The supplies shall be available in weather proof containers.

5.6.6 *Performance Guarantees*

Those given by different manufacturers shall be compared for maximum degree of conversion, minimum attrition loss, less susceptibility to impurities in feed streams, overload operations (permissible minimum/maximum operating temperatures), maximum feed rates or concentration of the reactants and minimum kindling temperature required. The guarantees shall be achievable in the existing plant design and equipments (heat exchangers, absorbers, etc.) being used.

Overall impact of using a particular catalyst on the operations:

- Should reduce the cost of production, corrosion of equipment and wastage of raw materials due to incomplete conversion using ineffective catalyst.

Chapter 6
Safety Management

6.1 Introduction

Many chemical plants operate with pressure vessels, high-temperature processes, and equipments where corrosive acids and alkalis are handled, and store toxic or inflammable substances stored in considerable quantities. Dangerous situations can develop during plant operations (or even when the plant is not operating—e.g. failure of cooling water spray on storage tanks of highly volatile and inflammable solvents can be dangerous even when the plant is not operating). A safe plant design improves employee morale and confidence, increases productivity, improves product quality, and can contribute significantly towards pollution control also. The profits and reputation of the company can also improve as the employees can be more attentive and the chances of mishaps are minimised.

Hence, production managers of all chemical industries make efforts to run their units safely and without any environmental pollution. Quality control and plant efficiency are very important and they look into these also.

6.2 Preliminary Analysis

Basic Engineering Document is prepared initially (or obtained from technology providers/designers who are engaged for this task) when a new industry is being set up, an existing unit is to be expanded or diversified with new products. It gives the product mix, process flow diagram, process description, information about utilities and raw materials required, the typical consumptions of raw materials and utilities, broad specifications of the various equipments, land required, preliminary plant layout, etc. Existing site infrastructure and local climatic conditions (ambient temperatures, rain fall) shall also be included in the document.

K.R. Golwalkar, *Production Management of Chemical Industries*,
DOI 10.1007/978-3-319-28253-4_6

The plant management shall make copies of the document and get them studied by senior experienced technical personnel in their organisation. If such persons are not available, the task can be done by appointing experienced consultants.

A working process flow sheet may be prepared wherein more details of the storages, material-handling methods, process units, an approximate layout, and operating conditions shall be entered if handling of corrosive, toxic, and inflammable materials and process units operating under high pressure/high temperature are involved.

Since there could be possibilities of fires, toxic gas release, and mishaps in the process plant, the following should be analysed (and discussed further) to identify the likely hazards (HAZID) and their sources in the scheme:

- Examination of the design of the process (heat and material balance, use of strong acids, inflammable solvents, etc.), piping and instrumentation proposed, facilities required for setting up the plant or for expansion, diversification, etc.
- Codes and standards followed for design and fabrication of plant equipments;
- Estimate the consequences of occurrence of hazardous events;
- Their probability of occurrence;
- Suggestions for reducing the hazards.

HAZID will thus create awareness at the design stage itself and can suggest need for suitable changes in the process design, modifications in equipment designs, and provision of additional safety devices as well as interlocks before the final designs are made. Any serious hazards identified must be addressed suitably to prevent accidents in future.

Examples

- A plant handling highly volatile organic solvents must have a refrigeration plant included in the Basic Engineering Document if it is going to be located at a place where ambient temperatures can go up to 45 °C.
- A reactor is operating with cooling water supplied from cooling towers for the jacket. An alternative source for cooling water supply shall also be provided. High-temperature alarms shall also be provided for exit water from the jacket.
- An air pollution control unit/effluent treatment plant shall be included in the plant design if it is not given in the Basic Engineering Document.

Such shortcomings can become clear during HAZID.

A report shall be made after the HAZID study and sent back to designer/technology supplier for making suitable changes. These can be related to:

- Design of process (possibility of operation at lower pressures/temperatures);
- Materials of construction (stainless steel instead of carbon steel);
- Equipment designs (more freeboard volumes, external cooling coils);
- Piping (pipes with more thick walls, higher pressure ratings, and arrangements to drain and flush them);
- Instrumentation (to monitor more parameters, with remote indication);
- Need for additional safety measures (safety valves, rupture discs, safety vents to release excess pressure);

- Precautionary measures (electrical safety interconnections for switching off drives for certain equipment) and warning alarms (for smoke and gas detectors and for high and low levels in process tanks);
- Provision of additional overhead water storage tanks for cooling which shall be available always—automatically and round the clock 24 × 7.

6.3 Revised Detailed Engineering Document

Revised Detailed Engineering Document shall now be prepared by them after making as many changes as are reasonable and practical. This revised document should incorporate the necessary changes in equipment design, fabrication details, piping and instrumentation diagrams (P and ID), plant layouts with locations of various equipments at different levels, reinforcing of supports, safety interlocks, electrical circuits, and operating conditions *for ensuring safety of personnel and plant equipments.*

This document shall again be studied by senior experienced persons very carefully in a critical manner from following points of view (before the equipments are fabricated, erected, and commissioned. *It thus saves costly changes which may have to be done later on*).

6.3.1 Process Details

Process details—The feed rates, concentrations, temperatures and pressures, and heat and mass balance for all streams and units shall be examined to check any major omissions or significant features.

6.3.2 Properties

Properties of all materials which will be handled (incoming raw materials, materials in process, and finished products)—their boiling points, flash points, toxic nature, and compatibility with each other (this is important for safe storage).

6.3.3 Procurement

Procurement of all such equipment which is likely to be hazardous during operation shall be done only after all statutory requirements have been complied with (their design, fabrication, testing, etc.).

6.3.4 Use of Guide Words

Use guide words such as no/less/excess/more/higher, etc., for various parameters such as flow of reactants, heating medium, cooling water, and levels in tanks for each section of the plant to examine further (more details are given below for such HAZOP study).

Some suggestions are given below for the reader.

6.4 Equipment Design and Construction

The study team shall look into the design, materials of construction, safety and corrosion allowances, fabrication, stress relieving, and testing of the process equipments (fired and unfired pressure vessels) before acceptance for using in the plant, *because they should be safe to operate.*

- Quality Assurance Plans shall be obtained from fabricators.
- Design and fabrication shall be as per international codes and standards such as DIN, JIS, ASME Sec VIII, IX, or any other prescribed by local statutory authorities.
- Test certificates for the material of construction (which must be suitable for the operating conditions) should be available. This should give details of chemical composition, tensile and compressive strengths, and limitations on use at certain temperatures.
- Corrosion-resistant material of construction shall be used and sufficient safety margins and corrosion allowance shall be added.
- Post-weld heat treatment (PWHT) for equipments must be carried out. Records of heating and cooling carried out during such treatments shall be available.
- Full radiography of welded joints shall be done as per statutory requirements.
- Non-destructive testing (**NDT**) such as dye penetration and hydraulic pressure tests are a must for at least 1.5 times the maximum operating pressure.
- All mountings and bought-out components shall be certified and tested (safety valves, gauge glasses, view glasses, and rupture discs shall be procured only with test certificates.).
- All statutory regulations and instructions given by competent authorities for inspection and procedures for various tests as above shall be strictly complied with by the fabricator/vendor.
- *An external limpet cooling coil provides reinforcement and hence is better than a jacket or an internal coil.*
- *Purchaser may not accept the equipment unless it meets all above conditions.*

6.5 Process Vessels and Machineries

All process vessels shall also be similarly studied for the details of their construction, inner protective linings if any, and their operating conditions. The study should consider the possibilities of accidents due to likely malfunctioning of new vessels.

Likewise, process machinery shall also be examined for materials of construction, in-built safety features such as internal safety valves (e.g. gear pumps), limit switches for overload, over hoisting for hoists, tripping of drive motor in case of jamming of lobes of twin lobe blowers due to dust, etc.

6.6 Structural Stability of the Chemical Plant

- Process vessels and their support legs shall be designed for maximum weight of the equipment and for dynamic load (at full speed of the agitators when agitated vessels are used) as if the vessel is full of material, along with the weight of fittings, gear-box, drive motors plus weight of connected pipes, vent pipes and for raw material storages on work platforms, weight of persons working there. In case of jacketed process units, the weight of heating/cooling fluid shall also be added by considering as if the jacket is also full. Wind load shall also be considered if the vessel is to be installed at a height. Civil foundations and buildings shall be designed for earthquake resistance (at least 7.0 on Reichter Scale) if dangerous materials are to be handled.
- Railings and Work Platforms shall have separate independent supports.
- Corrosion-resistant paint and protective enclosure by high-temperature-resistant cement, if necessary, shall be provided on the support members.
- All structural support members, i.e. vertical and horizontal beams, connecting braces of the building, and support structures for installing the process equipments shall be designed for maximum weight of the equipments with accessories along with the weights of work platforms around them, connected pipes, vent pipes, raw material storages on work platforms as well as the weight of persons working there (even though support legs of reactors may also be designed like wise). This is to ensure that even if some of the members get corroded due to any reason, the operations can continue till proper maintenance is carried out.
- All important equipment shall be well protected from rain, exposure to sun (if necessary), from dusty conditions, and strong winds blowing across by sheds and walls.
- Pipe racks shall be similarly designed for carrying the various pipes on them.

6.7 Design of Plant Layout

This shall be done while considering the safety of persons and plant equipments.

- Isolate storages for fuel and dangerous materials—these shall be away from main plant and shall have dyke walls and collection systems for spillages;
- Minimise chances of fire—(Main fuel tanks shall not be located near furnaces, HT electrical units, and ducts carrying high-temperature fluids.);
- Connect overflow nozzles of tanks for safe storage arrangement for fuels/acids/alkalis—in case of overfilling, the liquid from receiving tank will overflow into the other tank which is being used for consumption. There will not be any spillage. However, the correct amount consumed cannot be determined. The plant operating personnel shall be therefore careful, while a tank is being filled (Fig. 6.1).

6.7.1 Receiving of Raw Materials

- The weighbridge for incoming materials shall be near the entrance gate for keeping a record on all vehicles coming in and the quantities of materials being received.
- Railings shall be grouted at suitable places to prevent excessive movement of the vehicles during unloading—specially when they are reversed and emptied out automatically (e.g. preventing a truck loaded with sulphur falling in the underground melter pit when it is reversed and emptied in).

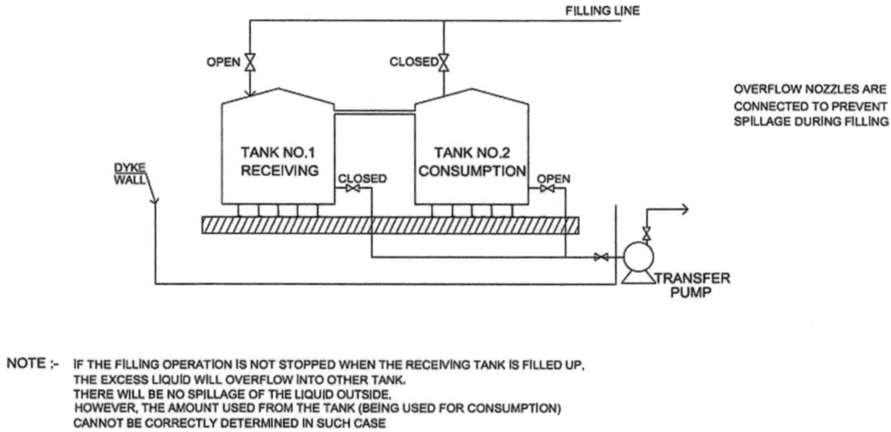

Fig. 6.1 Safe filling and storage arrangement

- Appropriate vacant space shall always be available to receive incoming supplies (separate bays in warehouses for non-compatible materials and empty storage tanks for receiving fuels, acids, solvents, alkali solutions).
- The exit from warehouse/stores shall never be blocked so that there is no mishap in the passage while going out (collisions between vehicles, slipping into storm water drains).
- Liquids from tankers can be received by connecting the drain valves to transfer pumps. The last remaining amounts of liquid in the tanker can be emptied out if the tanker is parked on a sloping floor towards the pump.
- Nitrogen flushing may be carried out if permitted—but do not over pressurise the tanker.
- Spillages from the tanker and leaks from pipes and valves may be collected in a small tank (with suitable lining) kept at ground level. It can be emptied out by a small submerged pump in the tank. It is not advisable to use compressed air to empty out the tank because it might have got corroded in humid weather and can burst due to the applied air pressure.
- Sufficient lighting, path indicators, and convex mirrors shall be available on walls for guidance of outgoing vehicles.

6.8 Other Precautions

- Location of process units and machineries on different floors shall have sufficient escape routes for persons working near dangerous units (pressure vessels, high-temperature reactors, and process units which can release gases from openings).
- Ease of operation—Control valves shall be easily accessible; work platforms shall be available at field instruments, near inspection windows and cleaning manholes for reactors, heat exchangers, and absorption towers.
- Working space shall be available around the process units for cleaning or for taking out any damaged internals, for maintenance of the unit.
- Escape routes, staircases, and railings shall also be provided at appropriate places.
- Alternate lights in plant shall be given supply from alternate power feeders. This will prevent total darkness in the plant in case of *power* failure of one feeder.
- There shall be easy approach to diesel/gas generator sets for the supply of power in case of need (*for emergency power supply*) when grid power fails. However, these sets shall start automatically whenever grid power supply fails.
- Provide dust extractors for crushers and grinders at appropriate locations with cyclones and bag filters.
- Provide induced draught fans and scrubbers for necessary equipments to suck out any toxic and inflammable vapours in working areas (e.g. electrolysis cells for chlorine/caustic soda plants).

- Water supply for important units shall be from more than one source, as well as from a dedicated water tank installed at a height.
- Installation of heavy and/or rotating equipments which can create vibrations shall be on the ground floor as far as possible since they can fall down if the support legs get corroded in the chemical plant (which can be possible). The vibrations can weaken the structures.
- Strengthen support columns and beams as per equipments installed on different floors.
- Location of plant control room shall be such that it is easy for keeping watch on critical equipments (feeding of raw materials, view glasses, level gauges, dial thermometers, and pressure gauges) even from a distance. These shall have shadow less lights with flameproof casings if inflammable vapours could be present.
- Release of vent gases from safety valves/rupture discs shall take place as much away from working areas as possible. The ends of the release pipes may be kept dipped in 10 % alkali solution if possible. This will minimise pollution due to release of gases.
- The loading points below overhead cranes or hoists shall be normally cordoned off so that no one can stand directly below them, while any load is being lifted up.
- The hoists shall have limit switches to arrest excessive hoisting, without which the wire ropes can break and the load can fall down.
- Fire fighting equipments for different types of fires (due to electrical, oil, reactants) shall be easily available at nearby.
- A dedicated water storage tank shall be available to immediately prime and supply the water pumps used for fire fighting.
- Gas detectors for toxic and inflammable gases (*please also refer to chapter on Instrumentation*) shall be installed near reactors, process vessels, flanged joints in pipes, and chemical storages.
- Routes of conveyor belts, pneumatic conveyors, and elevators shall not obstruct movement of men and equipments. Provide tripping switches along length of conveyor belts and elevators for use in emergency. These should get activated by pulling a string (laid parallel to the conveyor) from anywhere. Wide troughs shall be arranged below them for collection of any spillages.
- Floor shall be provided with tiles made from acid-/alkali-resistant materials and shall have anti-skid surfaces. There shall be suitable slopes towards storm water drains.
- Elevated storage reservoir shall be located centrally/near to as many process units as possible which need water always. The bottom outlet valve shall be operated by a solenoid control, and it shall open when power fails. It shall supply water to emergency use points such scrubbers, cooling jackets, electrode cooling, etc.
- Electrical equipments, motor control centres, and instrument panels shall be installed away from dusty areas (where crushing and grinding operations are

carried out) or from areas having evolution of corrosive vapours. These can cause short circuits and fires.

- Do not violate any statutory regulations—lifting devices, chain pulley blocks, and hoisting cranes shall be regularly tested as per instructions.
- Comply with all statutory instructions for electrical installations, fired and unfired pressure vessels, pressurised pipes, etc.
- **The readers can add more checkpoints to above list as per their own experience**.

6.9 Piping and Instrumentation Diagrams

This is one of the very important prerequisites for the HAZOP studies for the plant. The technical teams shall specifically study all details of the P and ID made for the gas and liquid ducts and piping in the plant:

- The fluids they will carry and their normal and maximum flow rates.
- Pressures, temperatures, suspended solids, and their solubilities (and erosive properties) since deposition of some solids inside can reduce life of pipes.
- Details of feeding system for materials into process reactors and vessels.
- Diameters of all pipes (and velocities of flowing fluids insides), **materials of construction, and their routes**.
- Design of expansion bellows and supports provided.
- Stress analysis reports.
- Provision of covers on flanged joints (the covers will protect persons from spray of fluids if there is a leak from the joint).
- Selection of flanges and gaskets as per operating conditions of the fluids (as above).
- Internal linings of ducts by refractory materials and expected surface temperature.
- External thermal insulations and claddings.
- Method of external heating (if required)—whether jacketed or by steam tracing.
- Examine the arrangements provided for releasing out vapours through a safe vent lines, flushing by nitrogen/water before opening for cleaning and maintenance.

6.10 HAZOP Study

Divide the plant in selected sections/parts for more detailed examination where a complete set of equipments (including accessories and machineries such as pumps, agitators, hoists, heating/cooling systems, and connected pipes) are installed and wherein a complete operation is carried out, e.g. grinding, absorption, and reaction stage. *If capacity of an existing plant is to be expanded or it is to be diversified to more products, the P and ID line diagrams shall be up dated first before carrying out the study.*

Make a detailed flow sheet for the individual sections showing all inputs, outputs, and *expected* process conditions (working of all parameters) for all process units and machineries as per P and ID. These shall be specially done for all pressure vessels and those process units carrying dangerous chemicals and their operating conditions.

Look into the operations to be carried out at high pressures/temperatures or where such conditions are likely to develop during operation; details of all feed systems, heating systems, and cooling systems are provided.

List all operating parameters such as flow of fluids, levels in various process vessels, temperatures, pressures, and concentrations of the chemicals.

Note the properties such as specific gravity, specific heats, pH, toxic nature, lethal dose, and boiling and flash points of all such chemicals.

(*Antidotes and immediate medical treatment recommended shall be known. First-aid facilities shall be available in or near the control room. All personnel shall be trained in first aid, and mock drills shall be carried out once a month*).

6.10.1 Further Analysis

All operating parameters and the events or situations such as excess feeding of materials into a reactor, high/low level in process tanks, high/low temperature of reactants, and cooling water supply less/stopped completely shall be considered.

- Guide words such as **no, reverse, very little, and excess** can be used to study all parameters. Examine the possible causes for the occurrence of such events human error, power failure, choking in pipelines, wrong indications by instrument, etc.
- Out of these identify which events can lead to the development of a hazard (such as overheating of reactants, tanks getting emptied).
- What can happen as a result? (loss of production, bad quality of products, and high pressure in reactors).
- Estimate consequences of the result—on population inside or outside premises, environmental pollution, damage to equipments, loss of production, etc.
- How often can such things really occur? Once in a while/any time/very rarely.
- What shall be done for safety purpose: (some typical precautions).
  - Advance warning by alarms (both visual and audio).
  - Provide thermostats for controlling heating by electrical means.
  - Provide positive displacement type blower instead of centrifugal blower where it is necessary to supply air for proper plant operation even when system is having some choke due to some reason.

- Pressure reducing valves can be provided to control steam pressure.
- Minimise occurrences by alternative arrangements (e.g. additional sources of cooling water).
- Provision of safety valves and rupture discs, automatically stopping electrical power to feed systems through interconnection with cooling systems, and shutting off fuel oil supply to burners if air supply is not enough.
- Strengthening of equipments to withstand such conditions if they will take place in spite of precautions. External reinforcements can be provided to minimise danger.
- Provisions of safety valves, rupture discs, and safety vents with the release of gases at a spots away from areas where persons may come.

Thus, apply the full study procedure when the design is complete and examine to ensure that all eventualities have been considered. This type of study on a large project may take a long time even many weeks and hence the options to be considered could be:

(a) To hold up, the detailed design and construction until the HAZOP study (with suggestions and clarifications from designers or till further corrections in the designs) are complete. *This can delay the project,* but will cost less since no actual/physical changes will have to be done or.
(b) Allow detailed design and construction to proceed and modify the detailed design or even alter the plant when the results of the HAZOP are known. *Apparently, the project is not delayed, but at the last moment, some changes may be required. The additional costs due to modifications could be high if major changes are required. In this case, there will be delay in the project also.*

6.11 Safe Erection and Mechanical Trials

All activities starting from dispatch of items from vendor's works till their trial runs shall be carefully looked into—especially for heavy items which can topple over and injure persons.

6.11.1 Dispatch from Fabricator's Works

All bought-out equipments and machineries shall be inspected in a stagewise manner during fabrication and also just before dispatch (which will be the final inspection). The loading onto the transport vehicle shall be carried out under supervision of the fabricator's engineer so that it is lifted and loaded properly using the lifting lugs and slings carefully. It shall be securely and safely tied to the vehicle and covered for protection from weather during transport. An experienced technician may accompany during transport till it reaches destination.

Safe routes shall be chosen for transport to the site, and the movement shall be reported to the sender and receiver every hour. **This is to minimise chances of any mishap on the way**.

6.11.2 Unloading at Site

Unloading at site may be done directly on the foundation if it is ready, or transferred to store and then shift to position as per erection plans. Weight of the equipment shall be known before attempting to unload. Arrangements shall be done for unloading by making available heavy-duty cranes, derricks, chain pulley blocks of a lifting capacity at least double that of equipment weight, and trained riggers. The area shall be cordoned off and no one shall stand in the way as the heavy item can swing suddenly in an unpredictable manner. Any temporary supports, platforms, and scaffoldings provided shall be strong enough.

6.11.3 Site Erection

Welding of supports shall not be half complete before loading in internals such as trays, tower packings, catalysts on trays, internal cooling coils, and provision of external fittings and connecting gas and liquid pipes.

No heavy piece of equipment or process unit shall be kept hanging, while persons go away for lunch or for any work.

No water tap, air blower, gas burner, etc., shall be kept running when there is no one around.

Internal lining of process units shall be done by experienced workmen under supervision of senior engineers only. The units shall remain isolated from all process units during this work. All connected pipes shall have blinds inserted in them—mere closing of valves is not enough. Fuses shall be removed from power supply lines to pumps. Breathing apparatus, escape ladders, warning bells, low-voltage lamps (not more than 24 V), etc., shall be available. The curing of brick linings shall also be carried out by experienced workmen under supervision of senior engineers only as it may require oil firing/acid brushing, etc. Exhaust fans shall be provided to suck out acidic fumes.

Similar precautions shall be taken during any internal work in a process vessel/unit.

All instructions given by erection engineers shall be followed.

Assistance from vendor and designers/consultants may be requested during erection, mechanical trials, and commissioning when special equipments are installed.

6.11.4 Mechanical Trials

There will be different procedures for the process units and machineries. These shall be studied and any doubt shall be clarified from senior engineers or the vendors. Some general precautions are as follows:

- Check proper fitting of foundation bolts of equipments;
- Gaskets of appropriate material and thickness to be provided in flanged joints. Provide covers on flanged joints;
- Confirm rating, direction of rotation, trip settings, and couplings of the motors for each driven unit to be taken trials of;
- Check free movement of rotating shafts by hand/wrench for bigger unit and quantity of grease/suitable lubricant in the bearings and gear boxes of rotating machines.
- Safety valves shall not be set at more than working pressure during trials.
- Do not feed in raw materials during trials. Check leaks of pipes and storage tanks by water and take trials of plant units by water/nitrogen gas/air whichever is safe and advised by vendor of the unit.
- Keep fire fighting arrangements and CO_2 cylinders ready at spot.

6.12 Safe Commissioning

Some general precautions are only presented here because there will be different procedures for different chemical plants as they will have a variety of process units and machineries.

Study the observations and results of the mechanical trials. Attend all leaks, any valve not operating smoothly, and any abnormal vibrations in machines such as compressors, and pumps blowers. Change the lubricating oil and grease as per instructions given in manuals.

Confirm working of all safety devices, safety interconnections, overload trip settings, working of effluent treatment plant (along with availability of chemicals for treatment); DG sets, cooling systems for reactors and condensers; fire fighting pumps, gas detectors, etc. Clean (flush) all the pipelines and ducts. Check all other equipments required for the particular plant. All instruments and process controllers shall be calibrated and tested.

It is important to start the plant strictly as per instructions given by designers/ consultants of the plant (after process units have been brought up to reaction conditions by preliminary heating/starting flow of cooling water as required).

All process units shall be watched carefully—for any leaks, distortion in ducts, tilting of units, abnormal vibrations, development of high pressure or temperature, excessive surface temperatures of units or pipes, strange smells in the plant, copious fumes from chimney, etc.

All these indicate that a mishap may occur. Reduce feeding rate of reactants, heating by hot gases or steam, speeds of agitators, and gas injection rates in such a case and report to senior engineers/designers/commissioning engineers deputed by vendors.

Draw samples from inlets and exits of process units to see whether the process is under control.

6.13 Standard Operating Procedures for Safe Working

After getting a clear confirmation that there is nothing wrong, one may proceed to step up the production rate slowly till the rated capacity is reached and all operations are stabilised. Fine-tuning of the plant units may now commence till best performance levels are achieved. All standard operating procedures and instructions for dos and don'ts are to be obtained from original equipment manufacturer (OEM) and discussed with senior engineers. The plant shall be run thereafter as per instructions.

Checklists of important observations required for safe, pollution-free, and efficient operations of the plant which are to be made every half an hour, every two hours, and every four hours shall be clearly written down in standing instruction book. Instructions shall also be available for checking all safety devices while taking charge of the plant from previous shift and while handing over charge to next shift.

Normally, no electrical interconnection provided for ensuring safety shall be bypassed. Any fault in these must be immediately attended.

6.14 Personal Safety Items

Face shields, hand gloves of full length, industrial safety shoes (with inner steel plate to protect foot and fingers) portable breathing apparatus, gas masks with appropriate gas adsorbent canisters attached, gum boots, industrial safety shoes, and acid-/alkali-resistant dress are some of the personal safety items. The time for which the gas canisters are used shall be recorded and they shall not be used beyond expiry of their capacity.

The dedicated water tank is generally an overhead tank which is always kept full by an independent source of water. It is connected to the safety showers and eyewash fountains. These shall be installed in areas where strong corrosive acids, alkalies, and other chemicals are being handled.

6.15 Ensuring Safety After Expansion of Capacity

A study is to be carried out to assess the maximum operating capacity of existing units and machines in the plant. This shall be compared with the required capacities for increasing the production rates. The shortcomings can be made up by modifying/upgrading the existing units or by incorporating new machines. Management shall train the production and maintenance personnel for the new process unit/machine being procured to ensure that they will be able to operate and maintain the new items without any mishap.

However, certain *undesirable changes in the operating procedures may get introduced inadvertently* during the course of operations for when expansion of capacity of the plant is carried out. There can be some wrong methods followed by the workforce for their own convenience—such as charging entire amounts of raw materials at a time and faster heating by using higher steam pressure (e.g. higher than originally instructed temperatures, concentrations, and feed rates to be maintained). Some of these changes may not be compatible with the process designs and *may result in operating the equipments or process units above their safe limits or may result in a dangerous situation unknowingly.*

The plant management must consider chances of mishaps from such methods or the short cuts adopted during operation of process units and the possibility of adverse effect on plant personnel, machinery, and also the surroundings as a result.

6.15.1 Plant Expansion

Plant expansion shall be carried out with more attention to safety since some of the existing systems may get overloaded such as conveyors for materials, refrigeration plants, electrical main bus bars, cables, incoming transformer, pressure vessels, heat exchangers, condensers, effluent treatment plants, and hoists even when operated carefully—because they may be already operating near their maximum capacity—*and this can create dangerous situations.*

Fresh HAZOP study shall be carried out to consider the possibilities of accidents due to likely overloading of existing units or malfunctioning of new machinery or *if the plant personnel do not properly understand the operation or maintenance procedures (chances of human error).*

Since an existing plant and new equipment can benefit from the study, a project may be studied many times in its lifetime say, and every six months (or earlier if feasible) after, the plant is in operation or whenever new equipment/machinery is being procured.

6.15.2 *Diversification to New Products*

- Examine the existing infrastructure—cooling systems, water treatment plants, ETP, availability of power, cogeneration facilities, packaged boilers, and DG sets?
- How many of the existing process units are not used or used only partially?
- Material and utilities requirements for new products—will they overload existing system? Can staggering of their running time be done to distribute the load more uniformly on motor control centres, electrical cables, hoists, DG sets, belt conveyors, Reactors, and cooling systems—if there are some common ingredients for old and new products?
- Will introduction of more products divert operator attention or need more maintenance effort?

6.16 Safe Maintenance Practices for the Plant

Plant engineers shall carry out condition monitoring of all key equipments. Preventive maintenance schedules shall be drawn and implemented. Sufficient inventory of critical spares shall always be available to avoid development of dangerous situations due to continuous running of equipments without repairs.

Regular inspections and hydraulic tests of all pressure vessels (boilers, economisers, air receivers, compressed air systems, gas cylinders, and all pressure vessels in refrigeration plants to be carried out), and load tests for all hoists, chain pulley blocks to be carried out as per **statutory requirements**.

Confirm that all safety interconnections are in order and alarms are operating at set values.

Follow the standard maintenance procedures for the process unit/machinery in case of a break down. *Since this will be different for the various units, only* ***general precautions*** *are given here*:

- This will confirm the isolation of the concerned unit after the contents are drained out and the unit is subsequently flushed by water, air, or nitrogen (whichever is suitable).
- Keep ready fire fighting equipments and a running water hose.
- The crew shall wear personal protective devices such as gas masks, face shields, hand gloves, protective industrial shoes boots, and protective aprons as per need. In case inflammable vapours are present, the shoes shall have rubber soles so that there is no sparking due to rubbing on hard floor tiles.
- No naked flame by gas cutter shall be used unless hot work permit is obtained.
- For site welding: earth connection shall not be taken from pipelines or process tanks. Separate lead cable shall be used.
- Do not use steel hammers if inflammable vapours are present.

6.16.1 Instructions for Safe Maintenance

Consult the maintenance manuals and OEM for the order and method in which the parts are to be opened, repaired, and then assembled again, e.g. impellers of pumps, agitators in reactors, and leakage from flanged joints in acid lines, etc.

- How the parts are to be tested after repairs and also for testing the entire assembly after the repaired part is refitted.
- Procedure for trial run of repaired parts, restarting individual units, and the entire plant.
- Special lifting/repair tools to be used during opening/shifting process equipments.
- Thorough cleaning of the area and the tools after repairs.
  Waste liquid generated during flushing, repairs, and testing is to be collected and recycled to process or treated in ETP.

6.16.2 Important Safety Precautions to Be Taken Before Working Inside a Closed Vessel

- Completely isolated from all sides, i.e. close all valves, put blinds in connecting piping, and remove fuses from starters of motors (for agitators, pumps);
- Flush out all contents thoroughly;
- Release all vapours (or suck them out);
- Test inside of vessel by toxic gas detector;
- Provide 24 V lamp (do not use higher voltage supply);
- Rope ladder to be hung inside;
- Portable breathing apparatus;
- Hand operated warning bell;
- Position another person outside to keep a watch on person working inside and rescue him immediately if required.

Case Study
Replacement of a broken agitator in a closed vessel (Fig. 6.2).

6.17 Some Examples from Industries

These are presented to the reader to indicate the practical manner in which their own process plants can be examined to make them safer. The list is only indicative, and the readers can add more examples from their knowledge and experience.

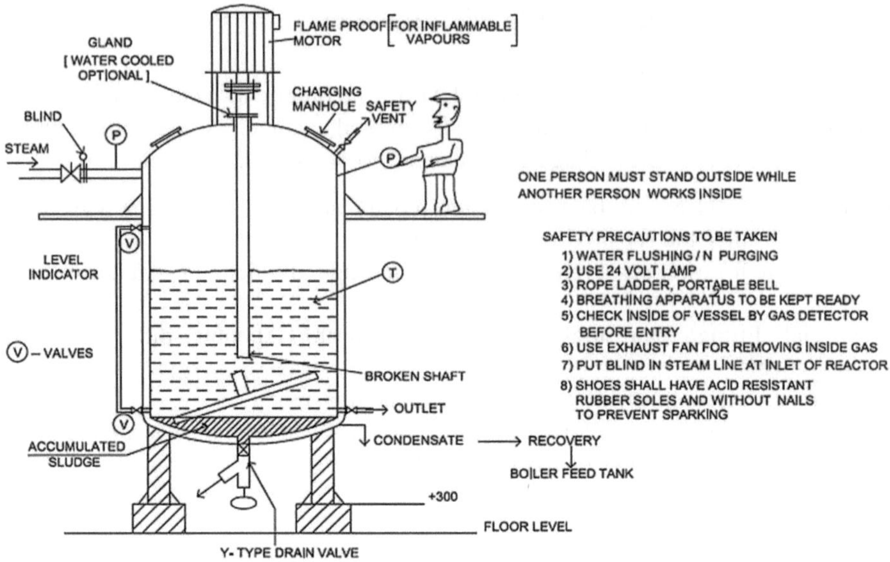

Fig. 6.2 Precautions to be taken while working inside a process vessel

6.17.1 Process Vessels with Heating Jackets or Coils

- Provide pressure reducing valves on steam lines to prevent higher pressure steam reaching the jacket or coils. Also provide safety valves on the jacket itself.
- Mechanical seals/water-cooled glands for agitator shafts.
- Flameproof motors if inflammable vapours can come out from the vessel.
- **Other process vessels/units**.
- Welded manhole covers instead of bolted covers (*which may leak at high pressure*).
- Provision of expansion joints in the internal refractory lining.
- Use metering pumps from small day tanks instead of feeding by centrifugal pumps from main tanks—excessive feed will not be possible. The day tank should not have a capacity more than 3–4 h operation.
- Use flowmeters and feed line of as small a diameter as possible to minimise chances of excessive feeding of dangerous liquids if metering pumps cannot be used.
- Provide double door arrangement while feeding solids/bags into the process vessels. Only one door shall open at a time. This will minimise ingress of air (if the vessel is operating at below atmospheric pressure) and also gases coming out (if it is operating above atmospheric pressure). This arrangement is useful for incineration of solid wastes (Fig. 6.3).

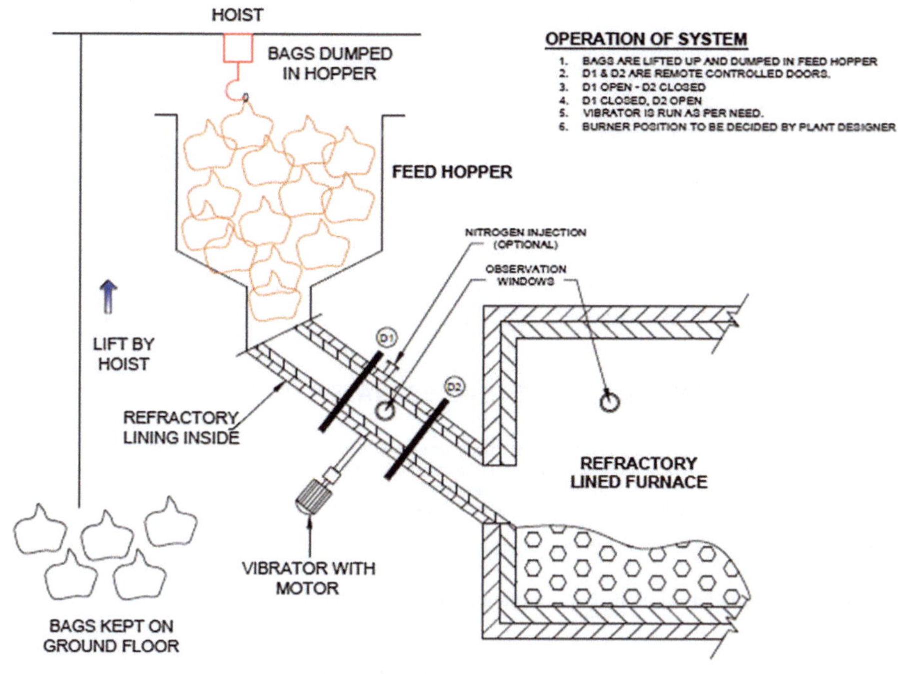

Fig. 6.3 Double door arrangement for feeding system

6.17.2 In Case of Vessels with Cooling Jacket

Provide a non-return valve on reactant feed lines to prevent high pressure developed in reactor exerting back on the feed tank (very high pressure may develop in reactor due to sudden high temperature if cooling water supply pump fails). This will prevent damage to feed tank. Reactor should have safety valves and rupture discs. Provide alarm if cooling water supply is inadequate. The alarm shall get activated by high exit water temperature and also if cooling water pressure is less at reactor inlet. Non-return valves may also be provided for cooling water inlet lines to prevent backflow of cooling water in case of supply pump failure.

6.17.3 Multiple Safety Devices

There should be a provision for more than one safety device (pressure and temperature switch, rupture discs, and safety valves) to ensure safety to persons and plant units even if one of the devices fails.

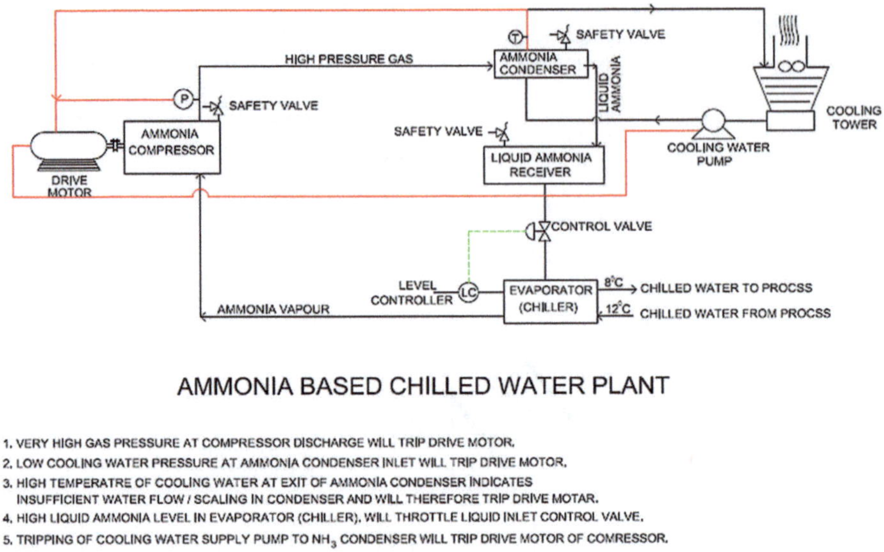

Fig. 6.4 Ammonia-based chilled water plant with safety interlocks

6.18 CASE Studies (Please See the Illustrations)

6.18.1 Ammonia-Based Chilled Water Plant

Ammonia-based chilled water plant has provision for multiple safety devices (pressure and temperature switch, and safety valves). These are shown in the Fig. 6.4.

6.18.2 Safety Interconnections for Oil Firing System

See Fig. 6.5.

6.18.3 Falling Film SO_3 Condenser

SO_3 reacts violently with water with evolution of heat. In case of a tube leakage, no water will enter on SO_3 side, because it is not under pressure. However, a little SO_3 can go to water side and make the exit water acidic and hence conducting. An alarm can indicate this. SO_3 vapours will get released to the sky from the top open end of the leaking tube (Fig. 6.6).

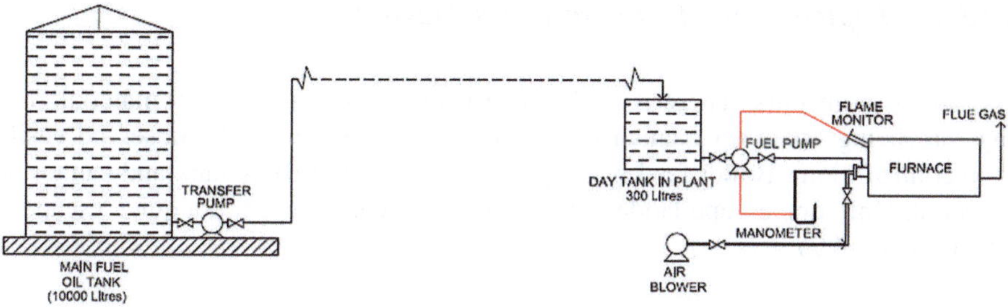

1. MAIN FUEL TANK IS LOCATED AWAY FROM THE FURNACE
2. DAY TANK IS LOCATED NEAR THE FURNACE (IN THE PLANT)
3. FUEL PUMP TRIPS IF A SIGNAL IS RECEIVED FROM FLAME MONITOR THAT FLAME IS EXTINGUISHED
4. FUEL PUMP TRIPS IF A SIGNAL IS RECEVED (FROM MANOMETER) THAT AIR PRESSURE TO BURNER IS TOO LOW.

Fig. 6.5 Safety interconnections for oil firing system

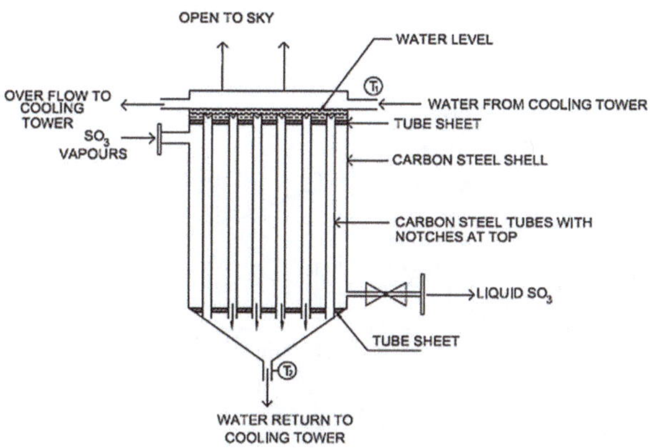

1. CONDENSATION OF SO_3 VAPOURS OCCUR ON OUTSIDE OF THE TUBES.
2. T_1 - TEMP. OF WATER AT INLET
3. T_2 - TEMP OF WATER AT EXIT

   IN CASE OF A LEAK ,NO WATER ENTERS IN THE SO_3 SIDE, BUT EXIT WATER CAN BECOME ACIDIC.
4. SO_3 VAPOURS WILL GET RELEASED FROM THE TOP OF THE LEAKING TUBE.

Fig. 6.6 Falling film SO_3 condenser

6.18.4 *Typical Unsafe Layout and Design*

The boiler is upstairs. There is no hoist to take up the bags of raw material. The bags are being taken up manually. Reactor and boiler cannot be observed easily from control room. Boiler feed water pump is behind the BFW tank and cannot be observed. Only one escape ladder (right hand side) can be used, but quick escape is difficult (Fig. 6.7).

6.18.5 *Improved (Safer) Fittings for a Process Reactor*

6.18.5.1 It Is Difficult and Dangerous to Remove the Accumulated Pasty Sludge

It is difficult and dangerous to remove the accumulated pasty sludge from the choked bottom nozzle of the reactor by simply opening the drain valve and inserting a rod from below. The sludge may suddenly come out and splash on the

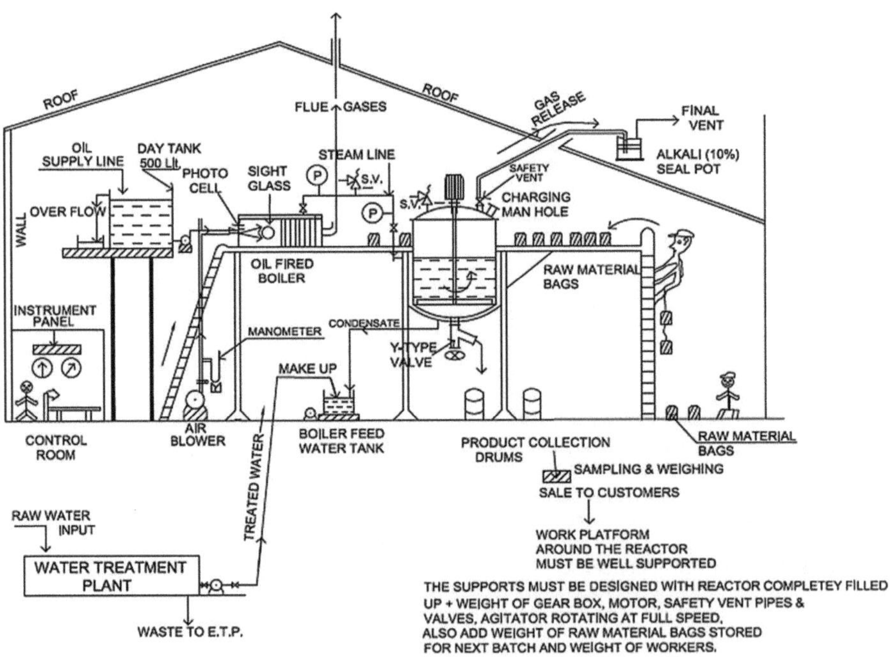

Fig. 6.7 Illustrative example for unsafe plant layout

body of the operator. However, the Y-type valve does not get choked as the plug is flush with the reactor bottom. On operating the wheel, the sludge can slide down and come out from the side opening.

6.18.5.2 The Wide Angle Observation Nozzle

The wide angle observation nozzle shown at the right allows a good view of reactor internals such as agitator, baffles, and movement of reaction mass. This is not possible with a narrow angle view nozzle shown at left (Fig. 6.8).

6.18.6 Pressurised Cooling System for Shell and Tube Cooler

This system can be dangerous if water enters the hot fluid side in case of a tube leak and reacts with it. If the water flow is under siphon, the hot fluid will enter water side and will come out from the open end (Fig. 6.9).

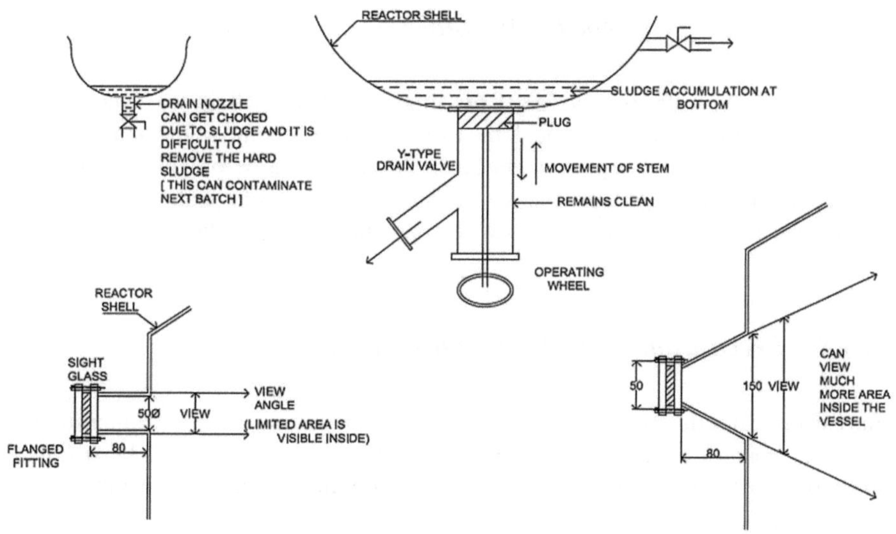

Fig. 6.8 Improved fittings for a process reactor

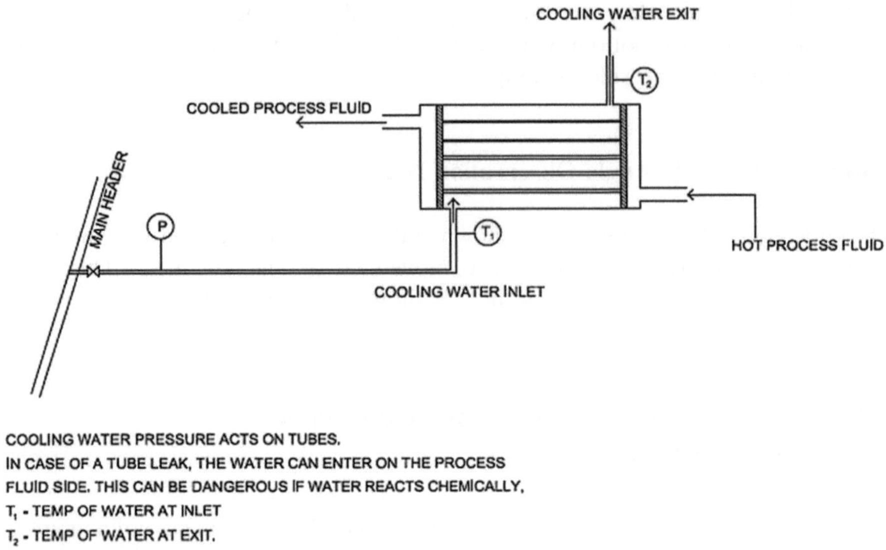

Fig. 6.9 Pressurised cooling system for shell and tube cooler

6.18.7 Cooling System Under Siphon

In case of a leak, water will not enter hot fluid side, but hot fluid will enter cooling water. If the hot fluid is a strong acid, the exit water will become acidic and its pH can be monitored in the pot kept at lower level. An alarm can be provided to warn the plant operator. The alarm can also be activated by the acidic exit water as it will become electrically conducting and can complete the circuit for the alarm (Fig. 6.10).

6.18.8 Precautions for Operation of Boiler

Many chemical industries have boilers for supply of steam—they are either fired by fuels like coal/oil or recover heat from hot process gases. It will be useful to familiarise all plant operating persons with the important points for smooth and safe boiler operations.

1. Water treatment plant shall be always regenerated in time to supply sufficient makeup water of specified quality to the boiler feed water tank.
2. Calibration of conductivity metre, TDS metre, etc., shall be checked regularly.
3. Level in boiler feed water tank shall be checked every hour. It is advisable to check working of the water treatment plant at the beginning of every shift.

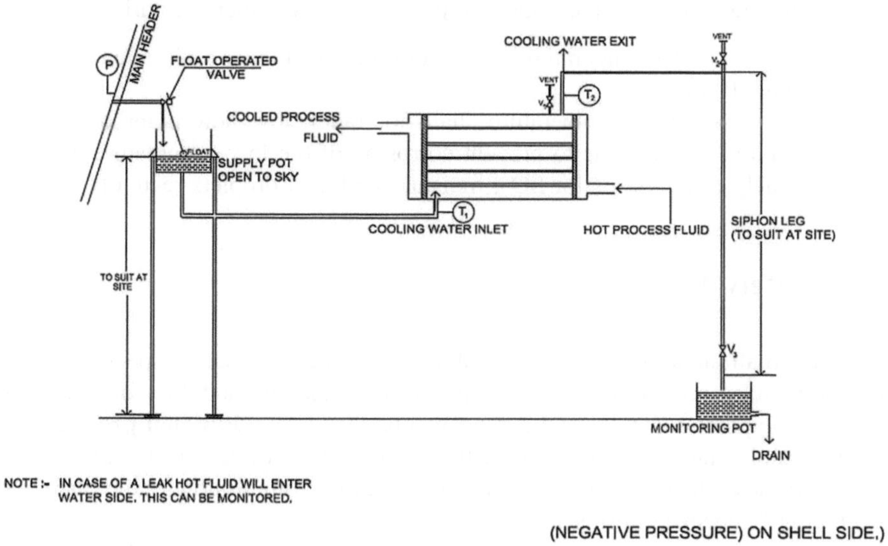

Fig. 6.10 Cooling system operating under siphon

4. Run boiler feed water pumps alternately so that all of them are in working conditions always.
5. Monitor water level in the boiler regularly. Gauge glasses shall be regularly cleaned.
6. Feed water quality must be analysed every 8 h and must meet specifications for TDS, pH, conductivity, etc.
7. Maximise condensate recovery from various points in the plant—specifically from steam turbine condenser and process units (in cogeneration plants).
8. Samples of water inside the boiler shall be regularly analysed for TDS, alkalinity, and blowdown quantity, and phosphate dosing and alkalinity shall be adjusted accordingly.
9. Check low-level alarms and very low-level tripping device at least once a week. When the water level goes very low, the tripping device shall trip the firing system/hot gases input to the boiler.
10. Operation of all valves for level gauges, safety valves, non-return valves, and feed control valves to be checked regularly.
11. Settings of safety valves shall never be tampered with. The initial setting shall be done by statutorily authorised boiler inspector only.
12. Monitor exit steam pressure and temperature at boiler outlet line regularly.
13. Cogeneration units—monitor inlet and outlet steam pressure, temperature and flow of steam and water through superheaters and de-superheaters regularly is to ensure proper steam quality at the inlet of steam turbine.

14. Monitor flue gas inlet and outlet temperature of the boiler regularly;
    (i) Flue gas inlet temperature shall not exceed the higher limit given by boiler manufacturer;
    (ii) Flue gas outlet temperature shall not fall below dew point of the gas mixture at any time to prevent corrosion of the low-temperature end;
    (iii) Analyse flue gases for SO_2, moisture and acid droplets regularly.

6.19 Safety Valves

Since the malfunctioning of safety valves on process vessels (especially those operating under pressure) can result in serious accidents and loss of lives and property, the plant engineers and managements must take **additional precautions**. Some possible causes for the malfunctioning of safety valves are therefore indicated, and these shall be looked into carefully before commissioning the plant and also regularly checked during running.

The plant management must get the safety valves, rupture discs, their installation, pressure release setting of these devices, vent lines, etc., inspected and approved by statutory inspecting authorities in their respective countries. There shall be full compliance with the instructions received from such authorities. **No existing statutory requirement on safety shall be violated**.

These are generally bought-out items from reputed and approved vendors. Vendor shall be asked to furnish test certificates from statutory authorities with the supply.

- Purchaser shall confirm the values for design, working, and test pressures from design engineer and statutory authorities *Setting of safety valves shall be done as directed by statutory authorities in their presence*;
- Should open in case of increase beyond set pressure;
- The valves shall be suitably sized to prevent further pressure rise;
- Close by itself after pressure is released;
- Shall provide tight shutoff during normal operations (to prevent loss of reactants through a leaking valve);
- At least two Safety Valves shall be provided;
- Reinforce the connecting nozzles for the safety valves with due consideration for the likely bending moment that may occur during blow off.

6.19.1 Markings on Safety Valves

- Name of manufacturer;
- Date of test;

- Smallest flow diameter upstream (shall be > 20 mm);
- Code–vapour/fluid;
- Set pressure values above normal working.

6.19.2 Malfunctioning of Safety Valves

Plant engineers shall take care of the following possibilities.

If the safety valve is fitted at the end of a long pipeline, it may not properly release the built up pressure inside the pressure vessel/process vessel because the long pipeline can have a high pressure drop or can get choked due to deposition of dust or ash (coming from the process vessel).

Hence, the safety valves shall be connected to the process (pressure) vessel by as short a pipe as possible so that it will not get choked due to dust/ash carry-over from the vessel (if the process is being carried out with a fluidised material)

- Bends should be avoided in the connecting pipeline to the safety valve.
- Release port must be of adequate size or greater than the required size. Seek professional advice for selection.
- High pressure drop due to choke in exit or inlet pipe from chemicals (e.g. *solidified sulphur deposits, salt deposits from boiling chemicals in the vessel*).
- Distortion of valve body due to weight of blow off pipe, over tightening of bolts.
- Too much blow off can create bending moment on connecting nozzle. *Proper supports and clamping required for inlet and exit pipes.*
- Corrosion and chattering of seat can lead to continuous blow off or a leak leading to loss of material.
- Jammed spring due to corrosion (in case of spring-loaded safety valve).
- Counterweight not fixed properly—sliding of counterweight can change the setting of pressure at which the safety valve should open.
- Hence, in most countries, it is recommended to use two safety valves.
- Do not depend on one safety device alone. Provide rupture disc in addition to safety valves. **Consult the local statutory authorities on safety and comply with their instructions.**
- The valves shall be tamper proof. Spring-loaded valves shall have protective covers on the springs so they do not get jammed due to corrosion, rains, dust, etc.

6.20 Preparing for Some Likely Emergency Situations

It is 3 or 4 AM in the night shift, raining heavily, main plant operator/engineer has gone to the toilet, power supply and plant lighting fail at this juncture, and there is a sudden process upset which may increase pressure or temperature in process unit.

Managements shall discuss every fortnight with senior engineers and experienced workmen some typical dangerous/problematic situations as given below *which may occur.* **This shall be discussed for each unit in the plant**.

Corrective actions should be planned in advance so that all are prepared to face such situations (Managements can invite suggestions if it is not possible to meet all persons.).

6.20.1 Points for Discussions

- What can go wrong in the plant? What can leak or break? What can catch fire?
- Can there be excessive temperature or pressure rise? Or both? Excessive gas release?
- Is it possible that uncontrolled feeding of some reactants/fuel/gases, etc., can occur?
- Can there be sudden rise (overflow)/fall in levels of process tanks?
- Can a sudden overflow or spillage of dangerous material (acid, fuel) take place?
- Failure of control valves (due to jamming of operating levers, failure of control circuit, and failure of compressed air supply).
- Sudden heavy rains, snow fall, or storm.
- Collapse of support structure of a process reactor/machinery.

In addition to above, the following shall also be discussed every fortnight or earlier:

- Design of electrical interconnections and tripping of drive motors in case of overload;
- Safety vents, valves, and rupture discs;
- Cooling systems for equipments and refrigeration systems;
- Suitable locations for safety showers and other personal safety devices;
- Fire alarms, high-pressure warning alarms;
- Limit switches for hoists, electrical overhead travelling (EOT) cranes;
- Provision of sufficient emergency lighting in escape routes shall be done.

6.21 Safety Organisation

6.21.1 Members (To Be Decided by Management)

- Safety Officer.
- Senior engineers shall be from each discipline, e.g. chemical (process operation), mechanical, and electrical (maintenance), project engineers, instrumentation, etc.

- Maintenance technicians.
- Representatives of workmen.
- Commercial and purchase officers to procure safety equipments and make special budgets for safety matters.
- Legal Officer.

6.21.2 Safety Policies

- Compliance with all statutory rules;
- Condition monitoring and preventive maintenance of key safety equipments;
- Immediate replacement of faulty equipments;
- Standardise operating procedures—revise if necessary;
- Prepare safety manuals for working on all equipments and machineries;
- Regular testing of all safety devices (safety vent valves, rupture discs, tripping systems, etc.);
- Ensure all workmen are given personal safety devices;
- Familiarise all persons with safety matters of their departments and hold regular safety drills;
- Train all personnel in fire fighting and first aid;
- Designate assembly point(s) in the premises and keep everyone informed;
- Display important safety instructions prominently.

6.21.3 Design a Work Permit System

Design a work permit system for carrying out any risky jobs.

- **Work Permits**
  Standard formats shall prepared for the process plant by senior engineers—these shall confirm the actions indicated but shall not be limited to the following list.
- Stop the reactions in the unit—which can generate heat, pressurise the contents, or generate toxic/inflammable vapours. Cooling operation may not be stopped (if it will not choke the nozzles due to solidification) well before opening it for maintenance.
- Completely isolate from all connected units from which reactants, water, gases, etc., can enter into the unit to be taken up for internal work. Put blinds in all pipes—only closing of valves will not be a guarantee for preventing such ingress.
- Remove fuses from motor starters—only switching off is not enough.
- These could be for cleaning a process tank having a dangerous material inside, taking out catalyst from converter passes, removing damaged internal

distribution trays or clogged tower packings, and repairing a broken agitator inside a vessel.

- List of the safety precautions to be taken shall be made by senior experienced engineers and plant operating workmen.
- It shall be complied with before carrying out such jobs.

6.21.4 Investigations and Corrections

- Record all mishaps and incidences of unsafe operations and working methods followed due to ignorance or carelessness by any one which could have resulted in mishaps.
- Categorise as per seriousness.
- Objectively investigate all mishaps and incidents.
- Establish reasons—human error, design fault, equipment failures (due to overloading or poor maintenance).
- Design preventive steps and discuss among senior experienced engineers and workforce.
- Submit for statutory approval (this may be required if an accident has occurred and the plant engineers have now designed a better system to prevent such accidents in future).
- Take appropriate corrective actions and confirm by trial runs, safety drills.

6.21.5 Follow-Up

- Revise the instructions and procedures if found necessary.
- Management shall allocate adequate funds under separate head of expenses.
- Review design of equipments, process conditions (pressure, temperatures, pH, etc.) every six months and whenever any major change is introduced in the plant.
- Review location of valves, sensors for instruments, and working of field instruments.
- Review ladders, railings, staircases, work platforms, and suitable lighting. Improve these for ease of plant operation and maintenance.
- Modify/change wherever necessary and implemented after approval by factory inspectors, electrical inspectors, etc.
- Maintain carefully and repair faulty unit immediately.

Chapter 7
Pollution Control

There are strict laws in all countries for pollution control which must be complied with.

It is realised by the modern managements that it is better to improve the working and process efficiency of the chemical plants so that generation of effluents can be minimised if not eliminated all together. Plant managers give very high priority to minimise environmental pollution by their production units. They aim for zero pollution by making efforts to eliminate the generation of all harmful effluents.

More efforts and emphasis shall be towards improving the process and technology, operational methods, and maintenance of the plant which is better than treating the effluents *after they are generated.* Some useful suggestions are presented here for considerations by the plant managements and all concerned persons for achieving this.

7.1 Reducing the Generation of Effluents

1. Selection of proper process and technology;
2. Proper design of the processing equipment;
3. Careful handling during transportation and storage of raw materials and finished products;
4. Adherence to standard operating procedures;
5. Condition monitoring and preventive maintenance of process equipment;
6. Expansion of production capacity and diversification to new products only after addressing safety and pollution issues;
7. Innovation of equipment and methods primarily aimed at minimising effluents;
8. Recovery and recycle of effluents as much as possible in own plant in order to minimise their quantities to be treated; and
9. Detailed study of physical and chemical properties of all materials being handled and their mutual compatibility.

K.R. Golwalkar, *Production Management of Chemical Industries*,
DOI 10.1007/978-3-319-28253-4_7

7.2 Economic Benefits of Pollution Control

- Better motivated workers who find it safer and easier to work on equipment; move about in the premises; and find it less noisy–smelly and less tiring.
- Reduced consumption of raw materials and other inputs such as power, water, and utilities per unit production of product. All this improves product quality and yields.
- Less accidents inside as well as outside premises.
- Less corrosion of foundations, and storage tank bottoms as well designed floors are made to drain away and dispose the spillages.
- Less downtime and better equipment life.
- A practice started for better coordination among marketing–production–procurement–maintenance departments for controlling wastes, which slowly develops into a habit. This results in smooth plant operation and hence less pollution ultimately.
- Cleaner and safer surrounding.
- Better company image/reputation—better goodwill for company.

7.3 Pre-operational Pollution

Pre-operational pollution—waste/effluents can get generated even before the manufacturing operations are started. Some typical reasons for this are as follows:

- Pretreatment of raw materials prior to feeding to the process units, e.g. crushing grinding (generation of dust), filtration (generation of filter cake);

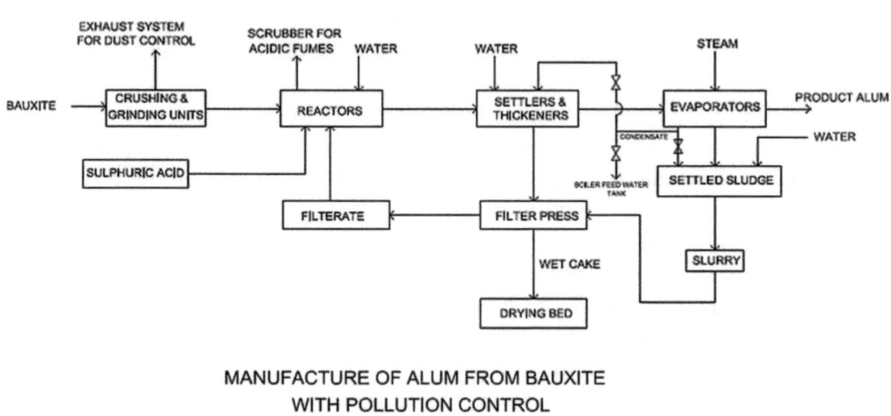

Fig. 7.1 Provision of exhaust system for the control of dust produced during crushing and grinding of bauxite before feeding to the reactors. It consists of cyclone and bag filter. The details of the equipment are given later on

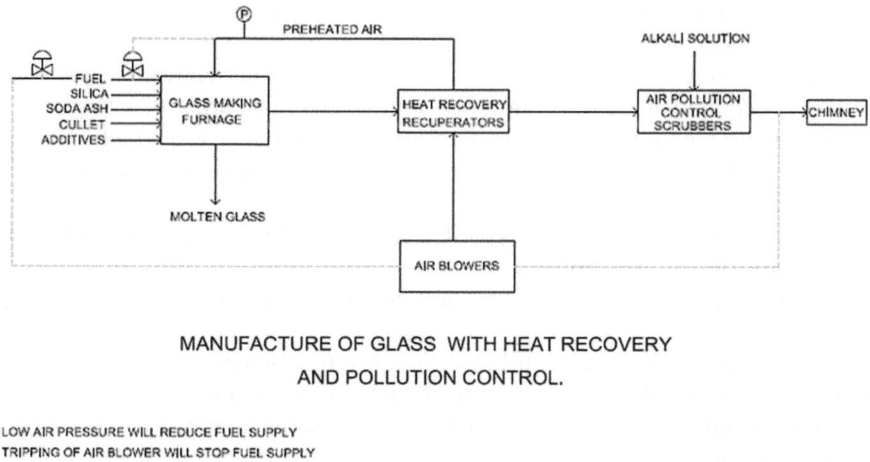

Fig. 7.2 The *green lines* indicate signals to air blower and to control valves in fuel lines. These are for automatic control of firing. The figure shows provision of scrubbing system with alkali solution as scrubbing liquor for exhaust gases from a glass manufacturing plant. The gases may contain SO_2 due to the presence of sulphur in furnace oil and use of sulphate additives with raw materials. A typical system has stack gas analyser which increases alkali flow if the gases have more SO_2 content

- Testing, washing, draining out the equipment that generates wastewaters;
- Occasionally, it is required to heat up the plant units to operating conditions (e.g. heating the catalyst beds to ignition temperatures prior to starting the plants) before raw materials are introduced into the process units. This can cause pollution due to burning of fuel/residual impurities inside; and
- Trial production runs of the plant during which effluents are generated (Figs. 7.1 and 7.2).

7.4 Some Typical Reasons for Environmental Pollution

7.4.1 Incorrect Design of Process Units

- Incorrect design of equipment (such as reactors, heat exchangers, condensers, coolers, grinders, dust collectors) which may cause incomplete conversions, inefficient heat transfer, and escape of volatile products or particulate matters. *Converters with too small diameter or absorbers with too less height have very less residence time for gases and can lead to inefficient operations.*
- Inefficient internals (less active catalyst and improper type of tower packing cannot ensure gas–liquid contact, a tilted distributor tray which will not irrigate

the tower packings uniformly and channelling may occur, improper cuts in baffle designs in heat exchangers will cause channelling on shell side);
- Incorrect choice of material of construction can result in frequent breakdowns of process equipment, leaks from process units, and puncturing of pipes;
- Not incorporating a filtration system for raw materials can partially mask the active surface of catalyst due to deposit of dust (which can affect conversion efficiency);
- Likely by passing of catalyst beds if a proper gas distributor is not provided in the converter at entry point of the bed;
- Channelling in absorption towers due to insufficient liquid feed distribution points;
- Tilted liquid feeders and distributors (not levelled properly) in absorbers;
- Faulty location of pressure or temperature sensors in idle pockets which can cause mistakes during manual operation or malfunctioning of controllers in process plants;
- Possibility of spillages of chemicals during handling in plant or transportation;
- Incomplete combustion of fuels due to fault in burner design; and
- Incorrect technology selection—use of only a packed tower for scrubbing instead of a proper combination of venturi scrubber and a packed tower with candle-type demister and an electrostatic precipitator.

The readers can add to the list and make it more exhaustive.

7.4.1.1 Fugitive Emissions

- These can occur from leaking flanged joints, seals of rotating shafts, glands of control valves, and nozzles for insertion of thermocouples or for connections to flow sensors.
- Hence care should be taken for providing (i) gaskets and flanges of appropriate materials (Teflon/neoprene rubber-/acid-resistant sheets) and necessary thickness, (ii) mechanical seals of shafts, (iii) sealing compounds at nozzles, etc. Chemical Engineers' Handbook shall be referred for selection.
- Suitable hoods and exhaust fans shall be provided to suck the fumes and gases from working areas, and these shall be properly treated by the scrubbing system in the plant.

7.4.2 Incorrect Operation and Improper Maintenance

- Sudden generation of effluents (*shock loads on the ETP*) during mechanical trials of equipment and machinery by not following careful methods for the trials.

- Frequently stopping and restarting the plant during commissioning stage itself leading to generation of off-spec products or process waste;
- Frequent product change over due to *pressure from marketing department*—requires frequent washing of process reactors and this can generate effluent streams;
- Plant overloading beyond rated capacity—can result in escape of unreacted inputs, acid mist, unabsorbed gases, etc.;
- Choked spray nozzles in scrubbers—these cannot deliver sufficient scrubbing liquor or can spray only on one side and cause channelling;
- Improper maintenance of key equipment such bag filters where damaged bags have not been replaced or delay in replacement of damaged spray nozzles in scrubbing towers;
- Malfunctioning instruments making process control difficult; and
- Wrong sampling procedures for process controls—giving a sense of complacency that process must be in control since all samples show results within set limits.

7.5 Preventive Actions for Pollution Control

7.5.1 Modify the Process

Minimise the escape of unreacted inputs from the plant by better process. A good example is the modified 3 + 2 DCDA process for the production of sulphuric acid wherein five catalyst beds are used instead of four. Three beds are used before the interpass absorption tower and two are after it. This results in the conversion of up to 99.85 % of SO_2 to SO_3. By providing separate acid circuit for the final absorption tower, the emission of SO_2 in exit gases can also be brought down further as compared to the earlier design where all acid towers had a common circulation tank.

7.5.2 Use of Better Catalyst

Use of better catalyst by loading caesium-promoted catalyst in the first and last passes of the converter, the conversion of SO_2 to SO_3 could be started at 390–395 °C in the first pass and higher equilibrium yield could be obtained in the last pass as it was operated at a lower temperature (400–405 °C).

7.5.3 Use Better Equipment

- Metering pumps instead of centrifugal pumps to prevent excess feeding of sulphur.
- Provide candle demisters in place of pad type for much better separation of acid mist.
- Provide flowmetres, concentration analysers, and temperature indicators at inlet of absorption towers.
- Horizontal furnace instead of vertical trickling burner type for minimising carry-over of unburnt particles of sulphur.
- Plate heat exchangers instead of trombone-type coolers for acid cooling were more efficient and had less breakdowns.

7.5.4 Equipment Design with Adequate Margin for Overload Operation

Process equipment must be designed for a certain overload, deactivation of catalyst, deterioration of performance of key equipment. Typically, following steps may be considered:

- Load about 15–20 % extra amount of catalyst.
- Heat exchangers shall have extra heat transfer area (however, provision of too many tubes can increase the cross-sectional area, reduce the gas velocity through the tubes, and reduce the heat transfer coefficient). Hence, one may increase the length of tubes instead of their numbers—this can increase the pressure drop, hence optimise the design by reworking various options.
- Provide circulation pumps for cooling water/scrubbers with about 20–50 % higher capacity.
- Provide additional equipment for pollution control.

An air pollution control system was designed for a waste disposal plant with venturi scrubber, two packed towers with cold water and alkali solution, an ESP, and an induced draught fan discharging through a 45-m-tall chimney.

*Another air pollution control system was designed for a fertiliser plant with wet cyclone, venturi scrubber, tangential spray tower and a packed tower with cold water (**and alkali solution—optional**), and an induced draught fan discharging through a tall chimney.*

Figure 7.3 shows two separate pollution control systems. The dust generated during crushing and grinding of the rock phosphate is arrested separately, and the acidic gases containing fluorine compounds are scrubbed separately. H_2SiF_6 solution is obtained from the system, and it is recycled to the reactor to reduce the consumption of fresh sulphuric acid. Part of the solution can be used to produce sodium silico fluoride as a by-product.

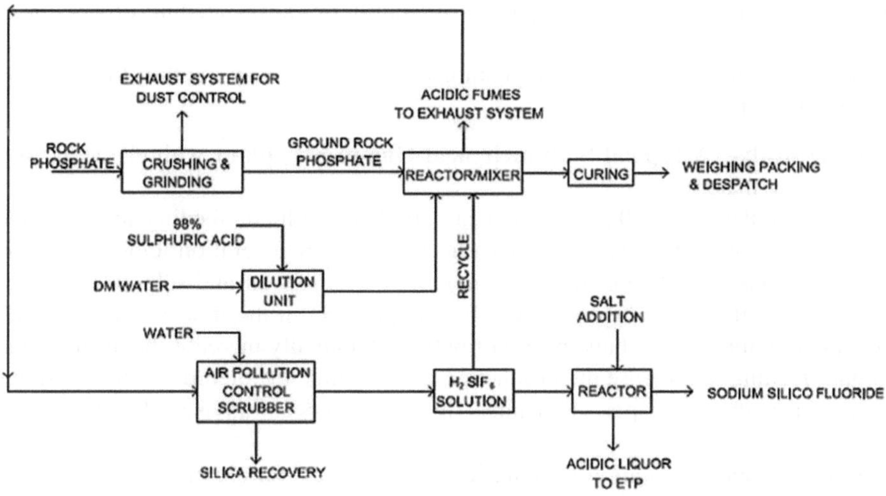

Fig. 7.3 Manufacture of single superphosphate with air pollution control

7.6 Standard Operating Procedures (SOPs)

Either obtain SOP from designers/technology suppliers or senior experienced engineers shall themselves prepare the instructions for starting, stopping, and running the plant. *Adhere to SOP and modifications, if any, worked out by senior operating engineers for reducing the generation of effluents*. Typically, these can be on the following lines.

7.6.1 Confirm the Following Before Starting a Process Plant

- Working of all safety devices, electrical interconnections, and diesel generator set for power supply during grid failure;
- Working of all the units of effluent treatment plant;
- Availability of essential chemicals for the ETP;
- Confirm the availability of storage space for off-spec products/arrangement for their disposal when the plant is being started for the first time, or after a major stoppage due to annual overhaul or after a major change in the plant;
- Working of all the units of refrigeration and cooling systems before starting heating;
- Availability of essential spares;
- Fill up seal pots at the end of vent pipes by alkali solutions;

- Take care to avoid wrong sequence of feeding or excess feeding or raw materials; and
- Maintain temperature and compositions of process streams within limits set by senior engineers.

Case study—A typical Sulphuric acid-Oleum-SO_3 plant with elemental sulphur as raw material.

Start circulation of alkali solution in scrubbers. Sulphur feeding to be started only when converter passes have been heated up to their ignition temperatures by suitable means (oil firing or other means). Start with low air and sulphur feed rates only. Bypass the waste heat recovery boilers partially in the first few hours till all the system temperatures have been normalised. Gradually increase the air flow and sulphur feeding to increase the production rate. Operate the oleum and SO_3 systems to produce them as per demanded by customers/as advised by marketing department.

Keep the acid circulation running in vent scrubbers on oleum tanks and liquid SO_3 tanks. This will arrest escape of SO_3 vapours always.

Stopping the plant: Raise the strength of circulating acid, system temperatures, and boiler pressure. Close dampers in gas and air lines to conserve heat.

7.7 Condition Monitoring and Preventive Maintenance of Key Equipment

Alkali circulation pumps regularly check the discharge pressure, flow through the scrubbing system, and return flow from scrubbers, and load on the motor, any abnormal vibrations.

Scrubbing tower and control system regularly inspect the spray nozzles, calibration of stack gas analyser, and the pH control system for scrubbing liquor, purging of spent liquor, and its treatment. *Attend even minor faults in alkali scrubbing system immediately.*

7.8 Contribution by Other Departments to Reduce Pollution

- Marketing department shall plan for reducing variety of different products at short intervals. Attempts shall be made to sell standardised products. This will reduce frequent changes in process conditions, and cleaning and washing of reactors in a batch production unit.
- Purchase department shall procure raw materials only as per specifications given by the production engineers—shall not procure raw materials with more ash,

more inerts, and more organics even though they may appear cheap as these can create more effluents. (Ref to procurement of raw materials).

7.8.1 Better Selection of Process Equipment

- Use root-type blower instead of centrifugal blower for getting steady air flow which is necessary for smooth plant operations.
- Higher capacity circulation pumps for better absorption of gases.
- Plate heat exchangers and cooling towers which can control the temperature of absorbing liquor in a better manner to minimise pollution (*instead of trombone-type coolers*).
- Install candle demisters instead of simple pad-type demisters for arresting acid mist.

7.8.2 Stage Inspections During Procurement

Sudden mechanical breakdowns of the units during regular production runs can create polluting streams in large amounts. Hence, it is necessary to carry out detailed inspections during fabrication, after final assembly (but before dispatch), on receipt at site and during erection as well as during mechanical trial runs prior to commissioning of the plant.

It is easier to carry out corrections before completing fabrication instead of attempting corrections after the start of manufacturing activities.

7.8.3 Procurement of Materials

Following must be considered during procurement to minimise pollution:

- The specifications shall be worked out to operate the plant in a smooth manner without compromising the product quality and safety of men and machines during handling and reducing equipment life. The raw materials shall be procured only as per specifications given by production engineers. Purchase department shall not procure raw materials with more moisture, more ash, more inert, and more organics even though they may be cheap. If the right type of material is not available or its delivery is delayed, the production engineers/maintenance engineers shall be consulted to see whether some of the specifications (purity, lump sizes, moisture content, the presence of some impurities) can be diluted *without causing pollution* (*or any other problem in the*

plant such as fouling of heat transfer surfaces by deposition of fine dust coming from the cheap dusty raw materials).
The presence of CaF_2 in rock phosphate results in emission of HF in the exit gases from superphosphate plants. It is a highly objectionable pollutant and is removed by multistage scrubbing of the exhaust gases. Excessive CaF_2 can overload the scrubbing system and may cause environmental pollution.

- Masking of active surfaces of catalyst can reduce the reaction rates or the conversion efficiency (conversion of input reactants to products). This can cause escape of unconverted reactants/feed materials with the exit streams or stack gases from the plant and can result in pollution.
- Excess dust from raw materials can cause choking of downstream equipment/reduced heat transfer efficiency/high-pressure drops in system/channelling in reaction (catalyst) beds which can result in escape of unreacted materials and can result in process problems such as generation of additional effluents.

7.8.3.1 Some More Precautions During Procurement and Storing

Well-defined specifications for containers for raw materials and packaging materials (bags, carbouys, liners in drums, etc.) make it easier to procure, unload, store them, and issue to operating units for production with minimum generation of effluents since spillages and wastage can be minimised during handling.

It is advisable to keep a storage tank empty always so that the fluid from a leaking container can be immediately transferred into it, thus preventing environmental disaster.

Methods of stacking bags, carbouys, and bottles (on sand layers, wooden pallets) shall be such that no breakages will occur during loading, transport (even on bad roads), and unloading at destination. The carbouys shall be of good quality tested by **stack tests before filling in**.

Inspection at entry point: Regular laboratory analysis of raw materials should be carried out to ensure smooth operations and minimum waste generation since they can more easily get converted to finished products, cause less choking in the plant units, and will not cause process problems (deactivate catalysts).

Stage inspection: It is also useful to analyse the intermediate products formed during the manufacture to keep a track of proper working of the plant and to find out at an early stage whether something is going wrong (rather than at the end of manufacturing operations).

Final checking of the products: It is also useful to ensure the saleability of the final product so that there are no stocks of off-spec products in the premises. Such materials will have to be purified or recycled into the plant. If this is not possible, they will have to be sold out at very low price or *disposed as effluents*.

7.9 Manpower Planning for Proper Plant Operation

- Qualified, well-trained, and motivated manpower is essential for the operation and maintenance of critical equipment such as boilers, pressure vessels, firing systems, refrigeration plants, filters, and distillation units. This can reduce the generation of effluents.
- **Automation for multiple process units**

  Manpower reduction can be planned only after considering the following:

  - Complexity of the process and the actual operations to be carried out,
  - Required frequency for monitoring the operation of process units,
  - Physical distances involved during working, and
  - Likely consequences of mistakes (loss to men and materials).

The required number of operators for the workstations may be planned for the plant on the basis of qualifications and experience necessary to carry out the process operations. Arranging refresher courses for revising the technical knowledge and regular training to correct any wrong methods also contributes towards better operation of the plant.

Proper ventilation and lighting of working areas make it easy to keep watch on plant operations and hence reduce the generation of effluents also.

Provision of sufficient safety devices and their easy accessibility increase the confidence level of the work force for taking up operation of complex plants.

7.10 Electrical Power

It is desirable to have diesel generators sets of sufficient capacity as stand by the source of power with automatic change over facility. These shall be used for operating equipment for emergency use such as alkali circulation pumps and induced draught fans for scrubbers, refrigeration units (*to prevent escape of volatile solvents*), fire hydrant pumps, boiler feed water pumps, and plant lighting.

The priority of the equipment shall be decided by the senior plant engineers.

Provision of uninterrupted power supply (UPS) arrangement may be done for critical units if there is a chance of escaping from very toxic or dangerous gases and chemicals.

7.11 Transportation of Materials

This aspect is very important for controlling accidents as well as pollution due to spillages of materials (raw materials and finished products).

7.11.1 Solids

These could be inert, inflammable (or explosive), hygroscopic, acidic, toxic, and self-igniting. These are packed in small bags (20–100 kgs per bag), bigger bags (500–1000 kgs per bag), or could be as loose lumps/powder filled in trucks or railway wagons. They shall be properly covered during transport to prevent contamination by dust or getting wet by rains.

Unloading at destination can be easier and faster if the loaded portion can be tilted by inbuilt mechanism (hydraulically operated system) in the truck. Spillage of solids can create dust pollution and corrosion of floors, and affect performance of blowers and pumps due to dust in the plant.

7.11.2 Liquids

These could be acidic, toxic, inflammable, with foul odour, viscous (difficult to pump), and with low boiling or flash points. Carrying such materials over long distances could give rise to pollution in case of an accident to the tanker or any spillage from a leak.

Avoid using the same tanker for many different products if possible. Considerable amount of liquid effluents are generated every time a tanker is washed to carry a different product. It is likely to contaminate the next product. A fleet of separate dedicated tankers is therefore preferable to carry different materials.

Plant engineers should consider the above properties of the materials which can result in the environmental pollution in case of a leak and accordingly equip the tankers/vehicles for transportation of such chemical liquids and solids.

Accessories for tankers could be fire extinguishers, pressure and temperature gauges, safety vents, lockable drain valves, earth connection by hanging chain, air compressor, spark arrestors, towing chains, and puller lugs. Sealing compounds, such as quick-setting cement and acid-resistant cement, shall be always kept in the driver's cabin for an emergency use.

7.11.3 Gases

Transportation of gases can be done by cylinders and special vehicles (including refrigerated ones) which are generally pressurised. Special precautions and permissions from statutory authorities are necessary before such transport. A leak can be very dangerous and environmentally highly polluting.

7.12 Design of Plant Layout

A properly designed layout of the process plant and the premises is very important in minimising pollution due to the following features:

- There shall be an ease of movement of vehicles (forklift truck, trolleys, etc) and less distances to be travelled internally for transport of materials reduce the chance of spillages. Convex mirrors and reflecting surfaces (with fluorescent paints) shall be installed at blind corners to prevent collisions between approaching vehicles and subsequent pollution due to heavy spillages.
- Safety devices such as safety showers, eyewash fountains, and gas detectors in adequate numbers shall be easily accessible, especially in areas that handling corrosive, toxic, and dangerous chemicals.
- Provision of anti-skid grating for work platforms can improve operator moral, and they become more attentive to process units.
- Provision of acid-/alkali-resistant tiles and water lines for washing sloping floors and tile-lined dyke walls around process units and tanks makes it easier to collect all liquids and *recover them if possible*. The spillages and floor washings which cannot be recovered shall be conveyed through suitable drains to effluent treatment plant.
- Dangerous, inflammable, and toxic materials shall be stored away from working areas to minimise the spread of fire and toxicity hazards.

7.13 Communications

Very clear instructions for plant operation and maintenance reduce mistakes by plant operating staff and maintenance crew. Plant working is improved, and there are less chance of wastage of materials due to better understanding of instructions and their actual implementation. Generation of effluents can also reduce. The exchange of information, suggestions, and instructions among senior production and maintenance engineers, junior workforce, marketing department, engineering stores, and external vendors thus improves the overall situation. More details are given in Chap. 14—General Administration (about communication).

7.14 Errors in Sampling Leading to Errors in Process Control

It is very important to get a truly representative sample of the untreated as well as treated effluents and analyse them for taking suitable corrective actions.

Solid wastes: small amounts of material are to be collected from various areas of the stored waste which can be mixed thoroughly to get a representative sample of the waste. Sufficiently large areas should be selected for this purpose.

Liquid effluents: the samples shall not be drawn from idle pockets in a flow channel, corner of a tank, or through a very long small diameter pipeline since the value of the parameter can change by the time the sample reaches the analysis point.

Exit gases: A continuous small volume of exit gases shall be drawn by sucking from the sampling point on the chimney, passed through a sample conditioner and analysed.

Hence, one should ensure this by minimising the time required for the fluid to travel in the sampling line, preconditioning before analysis is done (e.g. removal of acidic mist particles from an exit gas stream of a sulphuric acid plant being checked for SO_2 %), and drawing continuously from a flowing stream instead of drawing samples *intermittently* from the flowing streams.

It is possible that the sample is being taken out from the effluent/waste produced when everything is alright, while no sample is drawn when pollutant level is high in the effluents.

It can create a false sense of complacency that the pollution control units/ETP is functioning properly since the analysis of the treated sample may not indicate the presence of pollutants beyond permissible limits.

The online instruments are costly and difficult to maintain. One may choose manual methods for the analysis—but the frequency of analysis should be sufficient to draw correct inferences regarding functioning of pollution control units/ETP.

7.15 Standard Input and Output Norms (SIONs)

These consumption norms shall be well defined and compared with the actual consumptions in order to determine the wastages (generation of pollutants) or any unexplained loss of materials which could be due to the escape of reactants and/or products through gaseous or liquid emissions. A material balance of the plant can indicate the loss of unconverted reactants, intermediates, and finished products (which could not be collected) going into the effluents because in spite of the best efforts, there could be some generation of effluents. Inputs and outputs of all key process equipment shall be analysed and quantified if possible.

Apart from these, the other constituents of the effluents could be wastewaters from floor washing, spillages of chemicals, sludge drained during cleaning of reactors and other process units, backwash from filter press, spent scrubbing liquor, other solid waste from process units, filter cakes, spent catalysts and additives, etc.

Acid mist and gaseous pollutants form part of exit gases.

The entire plant may be run for about three to five days, in a manner as if a performance guarantee test is being carried out once every three to six months. This will indicate where things are going wrong.

7.15.1 Disposal of Effluents

Characteristics of the effluents (chlorides, sulphates, nitrates, etc.) heavy metals, calorific value, and ash content shall be studied to determine the disposal path. The guidelines given by Statutory Pollution Control Authorities in the country shall be followed for this purpose. The attached figures show typical pollution control schemes for solid, liquid, and gaseous effluents.

Following options may be considered for the disposal of solid and liquid effluents:

- Sale or disposal of the waste as such (i.e. without any treatment) if there is a buyer who can use the waste due to its heating value, the presence of free acidity or free alkalinity, or any other useful property.
- Recycle to own plant—to explore if this can be done (to recover some of the unreacted materials) without affecting the product quality and equipment life. *Filtrate from filter press in alum plants is recycled to the reactor; scrubbing liquor from HF scrubbers in superphosphate plants can be recycled to the rock phosphate and acid mixer for reducing the consumption of fresh acid.*
- Proper in-house treatment and disposal of the harmless material—this can be allowed by State Pollution Control Authorities, provided the treated effluent meets the prescribed standards.
- Dilution by addition of inert material (air injection in exit gases, addition of fly ash) and release only with a small further treatment—***this option is banned in all countries and shall never be used***.
- Sending to authorised secured landfill sites after rendering the effluent harmless by neutralising and immobilising the pollutants. This can be allowed by State Pollution Control Authorities, provided the disposed material in the landfill will not result in pollution of groundwater. Full details of the treated waste shall be given to the operator of the landfill site.

7.15.2 Disposal by CETP/CHWTSTD

The wastes which cannot be treated in-house may be sent to secured landfill sites or sent to Common Effluent Treatment Plants or [1]CHWTSTD facilities after carrying out primary treatment and entering into an agreement with the authorised operator of this activity and with permission of Statutory Pollution Control Authorities.

[1](Common Hazardous Waste Transport, Storage, Treatment and Disposal Facilities are operated by statutorily authorised agencies having properly designed vehicles and equipments for the transport and subsequent disposal arrangements. Certain fees are charged by them for the disposal depending on the characteristics of the waste).

These facilities are authorised in certain countries, and individual industries may be permitted to hand over their effluents **only after** the prescribed primary treatment.

Typical properties of the wastes which need careful primary treatment and safe packing/filling in containers are as follows:

- Solids—lumps, powders, sticky pastes, toxic, self-igniting, fine dusts;
- Liquids—viscous, toxic, self-igniting, acidic, explosive, low boiling/flash points, corrosive.

7.15.3 Disposal of Solid Waste

Only general guidelines are given here for disposal of solid wastes. The actual scheme shall be submitted to the statutory authorities in the country and implemented only as per their directives.

If the analysis of solid waste meets the guidelines given by State Pollution Control Authority (for chlorides, sulphates, phosphates, heavy metals) and has low calorific values, then it can be neutralised by the addition of lime, and immobilised by the addition of sodium silicate, fly ash, and portland cement as binding agent. It is now to be tested for oozing out of leachate when it is made wet (due to rainwater in practice), and can then be sent to the secured landfill site. This site is selected away from water bodies such as lakes, rivers, wells, and groundwater sources for drinking water.

- Provide layers of thick water proof membranes as liners at base and sides (to prevent migration of leachate or gas) made of high-density polyethylene at least 1.5 mm thick and tensile strength as per instructions from State Pollution Control Authority.
- Create cells at the site for disposal of only compatible wastes.
- Non-compatible wastes should be stored in separate cells.
- Leachate collection channels to a sump shall be made. Leachate removal by pumps or by gravity drains to holding tanks. Further treatment facility for the leachate shall also be made at site by physical, biological, or chemical methods.
- Dig monitoring wells in all four directions of the SLF and monitor quality of groundwater before and after construction of the disposal facility.
- Monitor evolution of gases from the site after it has been covered. Provision of gas collection and treatment facility shall be made from final top cover of the landfill site.
- Provide interceptor channels, drainage channels, and interceptors around the site. Those at the horizontalsurface of final cover shall have a proper slope for draining rainwater.

- The rainwater shall never get mixed with leachate collected even after closure period.
- Rectify the slopes to ensure proper surface run-off during operation and post-closure.
- Surface water drainage system—identify cracks, clean clogged drains, and storm water basin to prevent entry into the covered site.
- Provide an environmental monitoring system for air, surface water, and groundwater around the landfill site.
- Closure and post-closure plan will have to be made for long-term monitoring and submitted for approval.

7.16 Incineration of Solid Wastes

Managements of chemical industries are required to destroy the incinerable wastes generated in their plants by law as *per prevalent guidelines issued* by Statutory Pollution Control Authorities in the particular country. The facilities required will generally consist of installation of suitable incinerators, secondary combustion unit, air pollution control facilities, and exit gas chimney. Approval shall be obtained for these before commissioning their production plants.

Some typical guidelines are given below.

7.16.1 Waste Preparation and Feeding

Preparation of the feed in batches for the incinerator after checking the compatibility of components of the wastes and the concentration of pollutants (e.g. halogens, sulphurous compounds) in waste. The feed rate shall be controlled as per capacity of air pollution control facilities.

7.16.2 Considerations for Installing/Procuring an Incineration Plant

- The average calorific value and analysis of the waste (as above).
- The amount of waste which will be charged in per hour.
- Provision of double door arrangement for feeding of waste (please see Fig. 6.3 in Chap. 6).
- Design of the complete firing systems (with fuel tanks, feeding pumps, FD fan).

7.16.3 General Guidelines for the Construction of Incinerators Stationary Kiln/Rotary

- Main body shall be constructed of mild steel shells with refractory lining capable of withstanding 1300 °C, manual and automatic burners, instrumentation, safety valves for pressure release (auto and manual), and air blowers.
- Separate spray injection nozzles if liquid waste will also be incinerated some times.
- Ash removal arrangement.
- *Preferred fuel used shall be high speed diesel, liquefied petroleum gases, furnace oil.*

Rotary kiln-type furnace (if used)

- Carbon steel shell of adequate thickness (not less than 12 mms).
- High-temperature-resistant refractory lining (shall not be less than 45 % Al_2O_3).
- Furnace slope and rotational speed shall ensure a residence time for solid waste (up to 10 h) to ensure incineration.
- Inside operating temperature not less than 850 °C at start of feeding.
- *Safe arrangement for feeding waste.*

Secondary combustion chamber (thermal oxidiser)
The shell of the secondary combustion chamber should be provided with a refractory lining (capable of withstanding at least 1100 °C) in order to ensure complete destruction of the waste. This includes complete oxidation of dioxins and furans as well as of any material left unburnt in the rotary kiln.

- MOC and thickness of plates—not less than 12 mm;
- Refractory to withstand 1100 °C;
- Residence time shall be *minimum of 2.0* s;
- Temperature sensors at three places at least; and
- Photocell for flame monitoring.

Auxiliary burners
Incineration facilities must have at least one additional burner which should get switched on automatically when the temperature of the gases after the last injection of combustion air falls below specified temperature.

7.16.4 Air Pollution Control System

- Quencher for rapid quenching of exit gases from furnaces at 200 °C to prevent reformation of dioxin;
- A cyclone for particulate matter removal;

- Venturi and packed bed scrubber with alkali circulation for acidic gas removal;
- Demister pads or candles;
- Active carbon bed;
- ID fan of adequate capacity with rubber/fibreglass lining (for wet acidic gases);
- Tall chimney (carbon steel with rubber lining and a height of minimum 30 m) with sampling ports, corrosion-resistant inner lining; and
- Instruments for continuous monitoring of pressure, temperature, and stack gases.

General considerations

- Provision of emergency water and power supply (DG set);
- Inventory of fuel for five days consumption shall be available always; and
- Provide interlocking of feed system with exit gas analysis, flame control, and ***operating*** pressure in system (*shall be always negative to prevent escape of gases), tripping of scrubber circulation pump must stop ID fan and fresh feeding of waste. Safety pressure release valve shall open if there is positive pressure in the system.*
- Heat recovery—to be tried if possible.

7.17 Disposal of Liquid Effluents/Wastewaters

The scheme for liquid effluent/wastewater treatment shall be selected after an analysis of their representative samples and estimation of the volume. Please refer to the generalised scheme shown in the figure. An **equalisation tank** is provided to collect and mix the effluent so as to average out its characteristics. It also helps to minimise sudden shock loads on the ETP which can arise due to process upsets, washing of spillages, and draining out process vessels due to some reason. Various equipment and process units such as thickeners, lamella clarifiers, and tube settlers are important parts of the ETP. These are required for settling the sludge after flocculation and are available from vendors of ETP.

Dissolved air floatation and sequential batch reactors can also be considered.

Considerations for these important units for the treatment of **liquid effluent**:

- Flow rate as m^3/h;
- pH;
- Temperature in °C;
- Total dissolved solids;
- Concentration of suspended solids and their properties, density;
- Average particle size of the suspended solids;
- Presence of heavy metals;

- Presence of chlorides (*can cause corrosion of stainless steel parts*), sulphates, and nitrates; and
- Chemical and biological oxygen demand.

About five to ten litres of the effluent may be given to the vendor for carrying out treatability studies for designing the scheme for treatment (Fig. 7.4).

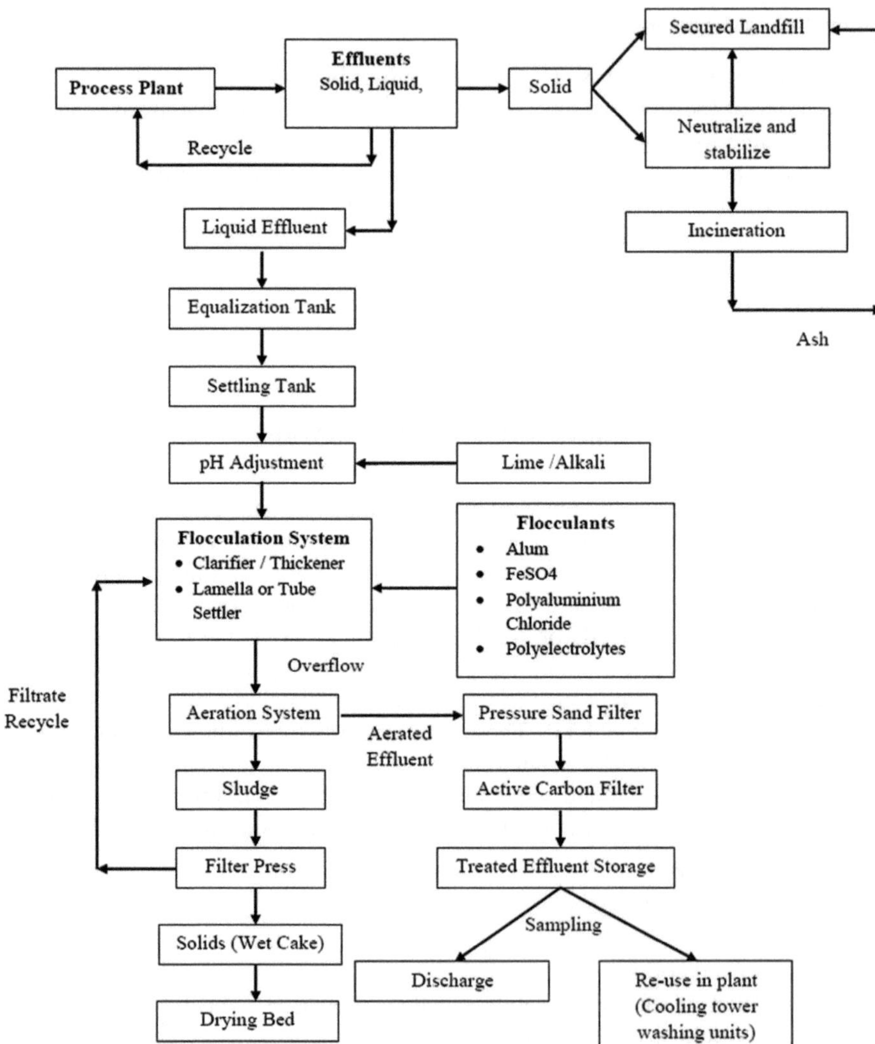

Fig. 7.4 Typical treatment scheme for liquid effluent/disposal of solid waste

7.18 Details to be Submitted by Vendor for the Approval of Plant Engineers

Vendor shall submit following details for the approval from plant engineers.

- Process flow diagram—including any pretreatment required for the effluents.
- General arrangement drawing of all process units.
- Guarantees for achieving the prescribed quality of the treated effluent (which must meet the standards set by Statutory Pollution Control Authorities wherein the pH, maximum TDS—total dissolved solid concentration, minimum dissolved oxygen, and maximum permissible concentration of heavy metals, chlorides, nitrates, and sulphates are given).
- Details of civil foundations and total space required for **each** of the treatment units.
- Empty and full weights for each unit.
- Power consumption KWH per cubic metre of treated effluent from the ETP (pumps, centrifuge, decanters, air blowers/compressors, and any other accessories included). Sufficient standby equipment shall be installed for the continuous operation of the ETP.
- Vendor shall furnish **test certificates for all pressure vessels** or the necessary tests shall be carried out by purchaser before acceptance. They shall be provided with standard safety relief valves.
- All vessels, valve, and pipes operating under pressure shall meet statutory requirements.
- Circulation pump capacity and discharge pressure, and motor with appropriate rating shall be checked.
- Arrangement to remove settled sludge/filter cake—*shall be available by non-choking drain valves (Y-type) and cleaning manholes for each unit.*
- Instruments provided for measuring and totalising flow rate of the incoming and outgoing effluent, pH, TDS content, and dissolved oxygen content of treated effluent from the ETP.
- Pressure sand filters and activated carbon filters shall also be supplied by the vendor along with the sufficient spare quantity of activated carbon.
- Residence time, location of inlet and exit nozzles, MOC of wetted parts in each process equipment, and drive motor details (AR bricks/stainless steel lining).
- Convenient and easy arrangement to take out internal parts such as rake mechanism, filter clothes, rotating parts and screen of centrifuge, aeration pipes, and nozzles shall be available if they get jammed in the settled sludge.
- Soil data (load-bearing capacity MT/m^2 area): to be given by the purchaser for the site where the unit will be located for designing the layout.

7.18.1 Removal of Sludge

- Quantity of sludge expected per day/per hour.
- Holding capacity of sludge chambers.
- Drain valve. Purchaser may ask non-choking (Y-type) valve to be provided by the vendor.
- Sludge removal pump capacity, MOC and type provided. Select with open impeller if necessary. Recycle line at sludge pump discharge may be provided to minimise choking of suction nozzle of the sludge pump.

7.19 Lamella Clarifier (an Important Unit in the Effluent Treatment Plant)

- Details of plate thickness, spacing between adjacent plates, angle of inclination, and length of plates in lamella clarifier to be informed by vendor.
- Level indicator shall also be available for water and settled sludge in process tanks.

7.20 Tube Settler

Tube settler can be used if it is difficult to take out and clean the plates of a lamella clarifier due to choking by sludge between the plates. Tube settlers may be found more convenient in such cases since the individual tube can be taken out, cleaned, and put back.

- Square tubes can be closely packed and hence can accommodate more number of tubes in a given cross section, as compared to circular tubes.
- Uniform flow of effluent in square tubes may be difficult if the size is more.
- Hexagonal and octagonal tubes can be better than square tubes as they have more internal surface area, and can be packed closer.
- Geometry of the arrangement and installation of tubes shall not disturb the settled sludge.
- Feed point location should be such that the settled sludge has minimum disturbance—to be informed by vendor.
- The feed distributor at bottom shall uniformly feed all tubes from bottom.
- Velocity of liquid in basin shall not exceed 0.9 m/min.
- Velocity of rising liquid in tubes *shall not* exceed 0.2 m/min.
- *Shape and angle of inclination of tubes to be indicated by vendor.*

7.21 Dissolved Air Flotation

These systems can be used for removal of suspended fine particles from the effluents.

Air is dissolved into water under pressure. (The solubility of air is more under pressure.) This is now mixed with the incoming waste stream, and the pressure is released in a separate chamber where by the dissolved air comes out of solution, producing a large number of very fine bubbles.

The air bubbles get attached to the suspended solids making them lighter and hence making them float to the surface. A skimmer at the surface removes these floating solids, thus cleaning the incoming effluent. It is then discharged from the system.

A small fraction of the outgoing treated effluent is taken and again saturated with air. This is recycled, mixed with the wastewater influent, and injected into the separation chamber where the pressure is again released to continue the process.

7.21.1 Consider for Selection After Examining the Important Components

- Treated effluent discharge tank;
- Pressurised Air saturation tank;
- Main froth producing tank (separation chamber);
- Mechanical skimmer;
- Recirculation pump;
- Air compressor;
- Piping and recycle control valves;
- Temperature indicators for incoming and outgoing streams;
- Level indicators for all tanks;
- Instrumentation—flowmetres for treated water and recycled water, level indicators for all tanks and separation chamber, dissolved oxygen metre for treated effluent (for special applications), and temperature indicators;
- Electrical panel for air compressor, recirculation pump, pressure gauges for compressed air and for discharge lines of recirculation pump, etc.;
- **Continuous DAF system may not need** the air saturation tank. The compressed air is injected in the recycle stream of the treated effluent. In this case, the control valve settings are done on basis of an almost constant characteristic of incoming effluent. This needs a sufficiently large equalisation tank and may also need chemical pretreatment by flocculants.
- Automatically operated control valves and bypass valve in the recirculation line with provision for manual override. **Vendor shall provide test certificates as per law for all units operating under pressure**.

7.21.2 *Disposal of Skimmed Out Wet Solid Material*

This is to be dewatered before disposal to reduce the sludge volume. This dewatering can be done by one of the following units/methods:

- Belt filter press;
- Centrifuge; and
- Drying bed.

Dried material can be sent to **authorised secured landfill site**.

7.22 Sequential Batch Reactor

- *All the treatment process is done within the same tank.*
- Primary sedimentation stage when the influent contains excessive suspended solids.
- Reactor basin.
- Waste sludge draw-off mechanism (by special sludge handling pumps or actuated valves).
- Aeration equipment.
- Effluent decanter.
- Process control system.

7.22.1 *Aeration Equipment and Air blowers*

Aeration equipment and air blowers are used *to provide aeration air* (for biological degradation of the organic components in the wastewater.)

- Power consumption for oxygen addition to wastewater per KWH consumed.
- Fine bubble aeration equipment shall be removable to facilitate maintenance of the diffuser nozzles.
- Dissolved oxygen probe within the basin for fine-tuning of the aeration system and controlling the aeration intensity.
- *Air flow control by motorised valves by carefully watching the periods of settling and decantation (no air supply during these).*

Inflow to the tank is normally interrupted during decantation, and hence, partitions may be made in the tank to continue the process and mixed liquor suspended solids shall not be washed out with the treated effluent. As a result, the total process cycle time can be reduced.

7.22.2 Effluent Removal Systems

The most important aspect of the design is to ensure that effluent is withdrawn uniformly from within the tank. Options to be looked into are as follows:

Fixed decanters including submerged outlet pipes with automated siphon control valves or a moving device and floating weirs-type outlet can be available. Vendor shall give the details in the GA drawing.

7.23 Air Pollution Control

Gaseous effluents could be harmful to vegetation; they could be toxic, acidic, or explosive, may spread over large areas causing respiratory problems to young and old persons by carrying dust, and may contain large concentrations of suspended solid/acidic liquid mist particles and inflammable vapours. The density of some effluent gases containing SO_3 may be high (as compared to air), and this can cause them to settle at ground level—especially during winter nights, or when wind velocities are low.

The acidic particles can be trapped by rainwater droplets and cause acidic rains—which is a major environmental hazard.

The gaseous effluents are to be treated in-house only and cannot be sent to the common treatment facilities. Various types of equipment and systems are available such as:

- Impingement separators, cyclone separators, graded bricks, gas filters, and bag filters;
- Venturi scrubbers, packed columns, ESP (dry and wet types), ejector scrubbers (Fig. 7.5).

7.23.1 Pollution Control System for Gaseous Effluents

Desirable features of the required system are as follows:

- All components should have very good corrosion resistance against dilute acid particles as well as an occasional droplet of stronger acid.
- Spray nozzles shall be able to cover the entire cross section of the absorber. The angle of spray shall be chosen accordingly. More number of nozzles may be provided to circulate sufficient scrubbing liquor for proper mass transfer.
- Arrangement to take out the arrested dust particles, while the system is running is preferred. If this is not possible, the frequency of cleaning shall be not more than once in about 30 days so that the process plant is not stopped frequently.
- Flowmetre for scrubbing liquor at inlet of scrubber.

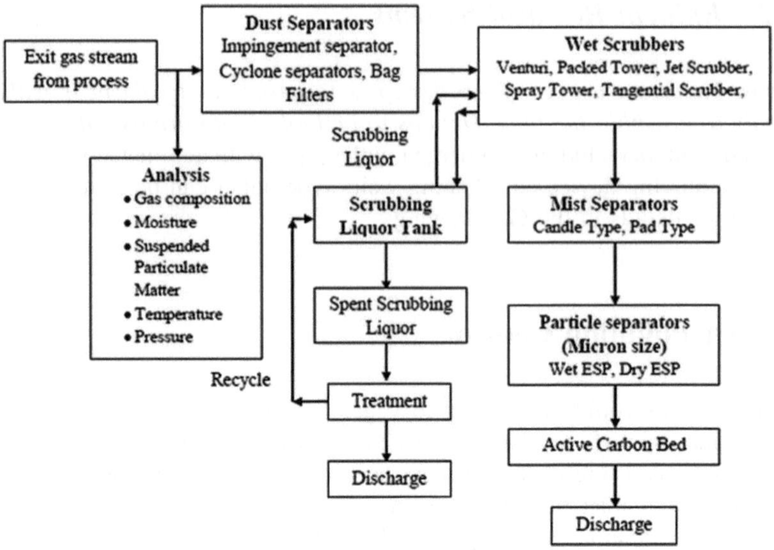

Fig. 7.5 Typical pollution control system for gaseous effluents

Example:

An interconnection of plant units with final absorption tower pump and scrubbing liquor pump (connection to alarm/hooter) in a sulphuric acid plant.

The air blower will trip if the acid circulation pump for final absorption tower trips.

The sulphur pump will trip if the air blower trips (Fig. 7.6).

Inflammable gases can be released through a specially designed Flare Tower (in petroleum refineries).

- Minimum pressure drop to flow of gases by selection of packings which can have high gas–liquid contact area as well as more void space per cubic metre of filled in volume.
- Ceramic partition rings with glazed surface can be used as supports for tower packing.
- Manholes shall be available for easy removal of internals such as spray nozzles and pipes and tower packings (single and double partition rings, tellerette rings) for cleaning and replacement if necessary.
- pH indicator and an automatic controller for maintaining alkalinity of scrubbing liquor for the treatment of acidic gases. Protection against high-temperature gases at entry point—this could be done by acid-resistant lining over which a temperature-resistant tile lining shall be provided.
- An alarm if the pH or level in circulation tank is too low.
- There shall be an installed standby circulation pump.

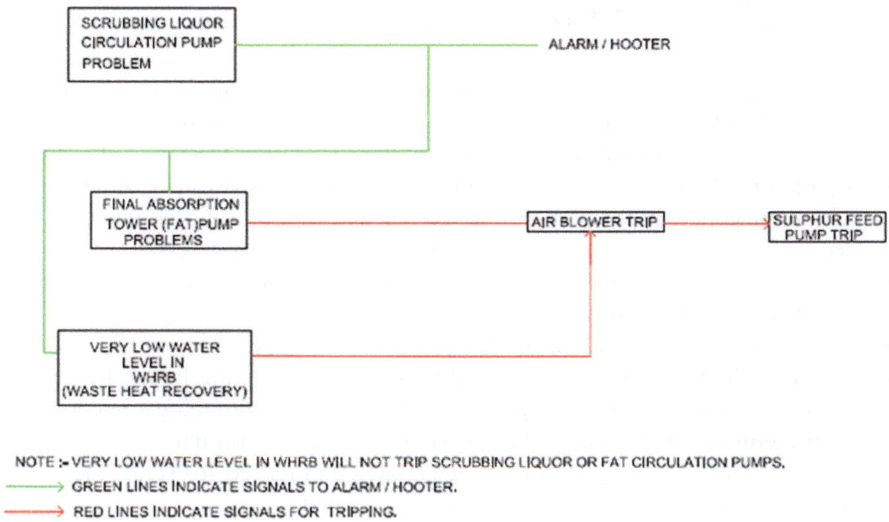

Fig. 7.6 Interconnections for air pollution control

- The bags in the bag filter can be of Teflon and shall be cleaned automatically by reverse air pressure applied through adjustable timer. They can get damaged at a gas stream temperature of more than 220 °C.
- The plant engineers/designers shall carefully estimate the normal as well as maximum flow of gaseous effluents to be treated and normal concentration as well as maximum concentration of pollutants.

7.23.2 *Inquiry for Gas Treatment/Scrubbing Systems*

- Maximum gas flow rate (nm^3/h) at present and in future;
- Composition of gas stream:acidic gases such as SO_2, HCl, HF, and the presence of toxic and inflammable gases;
- Particulate matter concentration mg/nm^3;
- Size distribution of particulate matter in the gas stream;
- Concentration and size distribution of acidic or dust particles;
- Temperature and pressure at which the gases will enter the treatment system;
- Moisture content in the gases (the wet acidic gases can damage the cotton fabric bags of bag filters. One may have to use costly PTFE bags in such cases. Higher moisture content can choke the bags if condensation occurs inside—hence operation shall be above dew point of the moist acidic gases).
- Dew point for the gas mixture.

7.23.3 Efficiency of Treatment Desired

- Maximum permissible concentrations of individual pollutant, e.g. acid mist, SO_2, HCl, NO_x, and dust particles in gases going out from the scrubber—*this must meet the standards set by State Statutory Pollution Control Authorities.*
- Separation efficiency desired for particles of different sizes to be informed by plant engineers (say for particles between 1 and 3 μ, 3–5 μ, microns etc.).
- Recovery of collected dust if possible (from cyclone, bag filters).
- Pressure drop on gas side during normal operation and at maximum gas flow rate shall be confirmed by vendor.
- Temperature of gases at exit of the system—by vendor.
- Products which are likely to be found in the spent scrubbing liquor–by vendor.
- Possibility of formation of suspended particles during scrubbing—vendor.
- **Arrangement for cleaning of the system—vendor to inform**.
- Can the spent scrubbing liquor be regenerated and reused for scrubbing?
- Power consumption per nm^3 of gases to be treated.
- Operating general arrangement drawing with scope of supply and exclusions from scope of supply—to be informed *by vendor*.
- Space required for installation.
- Expected consumptions of spares—by vendor.

7.23.4 Cyclone Separators

These are simple units and do not have moving parts. They separate the heavier dust particles in gas stream due to centrifugal force acting on the particles when it first moves down in a swirling motion in the unit. The direction of flow changes upwards on reaching the bottom, and the exit gas leave from the top.

Separation of the suspended fine particles takes place due to vortex motion of the gas stream.

Limitation of separation

The cyclones cannot remove all particles with satisfactory efficiency due to the generally wide variation in the dust particle size, weight, and properties, as well as variable gas flow rate in the process. However on the basis of known data, a design can be made and a major amount of dust can be separated. The author therefore suggests removable exit pipe pieces with different diameter and lengths. These can be fitted as a liner on the top nozzle.

Another cyclone may be provided at the exit gases in series to improve dust separation.

Vendor to inform the following

- General arrangement (GA) drawings showing gas entry and exit nozzles, overall dimensions, and empty weight. The filled up weight (with dust) shall be estimated by plant engineers for designing the support legs.
- Material of construction of the unit, location, and orientation of cleaning manholes, and maximum permissible pressure drop in the unit (at rated gas flow) shall be specified by plant engineers. Reinforcing ribs/flats provided externally.

Site jobs

- Sufficiently strong supports.
- Gas inlet duct piece for the cyclone inlet nozzle (may be supplied by vendor).
- External insulation and cladding if required, shall be provided at site.
- Arrangement for continuous removal of dust is preferable (instead of periodic manual removal) to prevent re-entry of settled dust in the exit gas stream. This can be a rotary air lock valve with a motorised drive. The motor shall have an instantaneous tripping arrangement if the valve gets jammed.

7.23.5 Bag Filters

- Bags of most materials can get damaged if the gases enter at above 210–220 °C. Consult manufacturer of bags before procurement.
- GA (general arrangement) drawings showing gas entry and exit nozzles, overall dimensions, and empty weight. The filled up weight (with dust) shall be estimated by plant engineers for designing the support legs.

Plant engineers shall check the following:

- Maximum gas stream temperature permitted by vendor;
- Pressure drop in clean bags system and when the bags are due for cleaning. In case of excessive pressure drop in the bags, a higher capacity ID fan may be needed or the vendor shall be asked to rework the design of bags;
- GA drawing and overall dimensions;
- Sampling points provided at entry and exit of bag filter;
- Ease of replacement of bags;
- Arrangement offered for cleaning of bags; and
- Accessories offered—air compressor, air receiver, adjustable timer, and control valves—shall be supplied by vendor.

7.23.6 Venturi Scrubber

Generally, venturi scrubber is used for the removal of particulate matter, acid mists, and similar pollutants. For efficient working, a very good contact between atomised scrubbing liquor and gases is necessary. Pressure drop through venturi scrubber is on higher side, and *hence, power consumption by gas blower can increase.*

A flow rate up to 2.0 l of scrubbing liquor/m^3 of gas flow may be required. Liquid injection spray nozzles can be at centre and in tangential positions.

The various sections of the venturi such as the converging part, throat, and diverging part shall be easily separable for cleaning. Hence, they shall be bolted together by flanged connections.

The removable throat piece shall have an adjustable damper to ensure proper gas liquid spray contact under different gas flow rates.

7.23.7 Packed Towers

Packed towers are used for further scrubbing after the venturi scrubber.

It shall be possible to introduce liquid injection spray at centre through pipes inserted in nozzles and in tangential positions for which suitable entry points (bigger nozzles for inserting the pipes) shall be already available on the unit shell.

- It should be possible to clean the spray pipes and nozzles in situ or to take them out completely—cleaning manholes and charging manholes shall be available.
- The MOC of spray pipes and nozzles shall suit process conditions of pH, temperature, and erosive nature of particles present, if any.
- Tower packing: Intalox saddles, tellerette rings, and pall rings may be considered.
- Self-cleaning packing: Polypropylene hollow balls. However, lower side manhole shall be available to remove them for cleaning.
- Top manhole shall be available to add more tower packings, to inspect internals, etc.
- Bottom portion of the spray tower/packed tower can work as circulation tank on which the tower will be mounted (this saves the footprint and a U-seal required at the liquid outlet of the tower).

7.23.8 Jet Scrubber/Ejector Scrubber

This is able to create a draught for the incoming gases due to the spray of scrubbing liquor. This will be useful for flow of gases.

However, liquid–gas contact is not as good as venturi because the mixing is not as intense. It is useful where concentration of pollutants and particulate matter is less.

The unit is mounted on a liquor circulation tank and thus saves separate liquid exit piping (with U-seal). The construction can be of fibreglass-reinforced plastic or rubber-lined steel tank.

Vendor shall be asked to supply a spray nozzle with a wide enough 135° spray angle so that the cross section is properly covered.

7.23.8.1 Spray Tower

This design of the tower does not have any packing, but has multiple spray pipes to cover the entire cross section of the tower. It is an empty cylindrical vessel, has lower pressure drop, and can be used for an initial treatment to partially remove pollutants.

7.23.8.2 Tangential Scrubbing Tower

It has multiple spray pipes or nozzles installed tangentially at more than one level. The tangential sprays clean the tower inner walls continuously and do not allow build up of layers of deposits on the walls of the tower.

These are used in phosphatic fertiliser industries where silica/other particulate matter tend to deposit on the walls of the tower. Polypropylene and HDPE can be considered as MOC. The pipes and valves shall be pressure tested before installation.

7.23.9 Tower Packing Items

These are used to increase the contact area of absorber liquid and the gases. Various types of tower packing items are available, e.g. plain Raschig rings, single partition, double partition, with outside corrugations, with or without glazed surfaces, and also available are tellerette packing, pall rings etc.

Tower packings shall be selected after considering the chemical resistance to scrubbing liquor at the actual operating conditions of maximum temperature and concentration; the surface area and void fraction per unit volume when dumped randomly in the towers; the expected pressure drop in the tower; and the actual crushing strength of the tower packings.

7.23.10 *Spray Nozzles*

These are important for irrigating the entire cross section of absorbers for proper gas–liquid contact in a pollution control units.

Selection criteria include the following:

- Properties of scrubbing liquid to be sprayed: pH, density, boiling point, viscosity, etc.
- Flow rate minimum and maximum as litres per minute per nozzle.
- Physical and chemical resistance to scrubbing liquor at operating conditions and the presence of erosive suspended solids in liquid.
- It shall be easy to remove, clean, or replace them.

7.23.11 *Entrainment Separators and Candle Demisters*

These are generally available as stainless steel or glass fibre pads or as candles (special fibres enclosed in cage) and are installed for arresting the liquid droplets entrained in the exit gas streams of equipment in order to reduce atmospheric pollution.

Selection criteria

- Corrosion and erosion resistance to gas flow at operating conditions.
- Maximum concentration of liquid droplets in gas stream, the particle size of liquid mist droplets, and their loading as mg/nm^3 and the maximum gas flow rate.
- Removal efficiency desired for various mist particles: plant engineers shall specify this. It depends on the harmful effect of the mist on downstream equipment, emission through chimney. Example: *should specify 99.0 % for all particles above 3 μ, 99.9 % for all particles above 5 μ.*
- Allowable pressure drop through the demister—*candle demisters have higher pressure drops (for the same gas flow) due to compact packing, but also have greater efficiency of separation.*
- Overall size of each pad or candle and the assembly proposed by vendor to achieve the required performance for separation of mist from the gas stream. More than one candle may be required. General assembly drawing shall be available.
- Very thin stainless steel wires of pads can break easily as they can get corroded due to the presence of dilute acid mist.
- The pressure drop on gas side shall be as less as possible as it can increase power consumption of blower. More than one pad is generally required; these are arranged about three pipe diameters apart, one above the other in the gas stream.
- A separate vessel for housing such pads or candles may be required or they can be installed by extending the height of the absorber. They shall be firmly fixed since high gas flow can displace them.

- Plant engineers to indicate their preference for the direction of flow of gases (laden with mist) through the candle demisters, i.e. whether outside-in or inside-out. This will determine fitting of candles on tube sheet). *Outside → in offers more area for filtration as compared to inside → out.*
- Seal pots shall be installed inside the absorber to collect the mist droplets from bottom of the candles and drain them by themselves into the absorber.

7.23.12 General

Electrical tripping interconnection shall be provided to stop the gas flow/trip the blower/stop feeding the raw materials/firing in the furnace if the scrubbing liquor pump fails to deliver due to any reasons. Audio visual alarms may also be provided if the emission of pollutants is more from the chimney.

7.23.13 Electrostatic Precipitators (ESPs)

These are installed for removing fine suspended particulate matter from gas streams which are otherwise difficult to remove by scrubbers, bag filters, dry and wet cyclone, etc. Electrical power at high voltage is used for charging the particles which get attracted towards the grounded collecting electrodes. The gas streams flow through the passage formed by the discharge and collecting electrodes.

Plate precipitators—old designs
Particles are collected on flat, parallel surfaces that are about 250 mms apart, with the discharge electrodes installed between the adjacent plates. The contaminated gases pass through the passage between the plates, and the particles become charged and adhere to the collection plates.

7.23.13.1 Tubular Precipitators

Working principle: tubular precipitators consist of tubular collection electrodes with discharge electrodes along the centre lines of the tubes. The dust laden gases flow through the tubes along the discharge electrodes. The dust particles get charged and are collected on the grounded walls of the tubes.

Factors affecting the efficiency of electrostatic precipitators:

Larger collection surface area and lower gas flow rates increase efficiency because of the increased residence time available for electrical charging and subsequent collection of the dust particles.

An increase in the dust particle migration velocity to the collecting electrodes increases efficiency. The migration velocity can be increased by:

- Increasing the gas temperature—provided it will not damage the internals.
- Increasing the voltage field.

7.23.13.2 Dry Electrostatic Precipitators ESP

Following data shall be available for selecting an ESP.

- Volume, pressure at which the gases will enter the unit and temperature of gases to be treated, concentration of corrosive gases such as SO_2, HCl, moisture content in gases, content of suspended particles mg/Nm^3 size distribution of these particles. These data must be provided to the vendor.
- Maximum corona power that can be supplied by each electrode—by vendor.
- Material of construction of the electrodes (by vendor).
- Power consumption of the ESP for the given operating conditions.
- Geometry of gas passages through the units—by vendor.
- Orientations and location of inlet and exit nozzles.
- Arrangement for the removal of settled/collected dust particles.
- Ease of replacement of damaged electrodes.
- Collection efficiency desired by plant engineers and offered by vendor for the particles as per size distribution mentioned in inquiry.
- Electrical power control mechanism and protection provided at high-voltage circuit.
- Pressure drop through the ESP when clean and when loaded with dust.

7.23.13.3 Wet Electrostatic Precipitators WESP

It is practically very difficult to achieve a high degree of efficiency by the dry ESP for the removal of the suspended particles in the gas stream due to their different conductivity. Wet ESP has a fine fogging spray of water at the gas inlet point to make the particles heavy and sufficiently conducting. The discs fitted on the discharge electrodes have many sharp points; hence, they can supply good corona power to the suspended particles. They are now attracted towards the collecting electrode tubes and remain stuck there (unlike dry ESP wherein they may re-enter the gas stream). Periodic washing of the collecting tubes can be done to remove the deposited particles to keep the active surfaces clean.

- This can keep the system pressure drop on lower side.
- These features can make the wet ESP work better than the dry ESP.
- A filtration system for the circulation water maintains the system clean.

- The casing, gas entry nozzle, and gas distributor can be in rubber-lined carbon steel. The electrodes can be corrosion-resistant alloys since wet corrosive gases will be flowing across them.

Details of electrodes, power control system, instrumentation, and details of arrangement to suspend the electrodes should be given by the vendor.

Unit should be well protected from heavy rains due to high-tension electrical fittings.

7.24 Chimney for Final Discharge

Site jobs are as follows:
Erection of prefabricated chimney procured from vendor OR.
Fabrication of large diameter chimney at site.
Following shall be considered.

- Maximum quantities of pollutants that may be released during plant upsets or during start up and the maximum gas flow rate nm^3/h through chimney.
- Composition of gases (quantity of pollutants such as SO_2, HCl in kg/hr).
- Temperature and pressure of gases at inlet.
- Whether self-supported chimney can be erected or will it need guy ropes supports. The ropes shall not create any obstruction for any movement of men or materials.
- Bottom plates shall be of appropriate thickness and foundation bolts shall be of high tensile strength so that the chimney will not topple over due to strong winds.
- Standard fittings and provisions to be provided: aviation warning lamps, lightening arresters (with copper conductors), safety ladders, gas sampling points, platforms, and railings. These are required in most of the countries.
- Protective lining (AR bricks) at bottom chamber and (rubber lining) along the vertical length since there could be moist acidic gas from scrubbers once in a while.
- Final Stack height may be estimated by:

$$H = 14\sqrt[3]{Q}$$

where H = stack height and Q = kg of SO_2 released per hour. Consult the local Statutory Pollution Control Authorities and get their approval.
However, the final height shall be decided after considering the following points:

- Location of plant and local climatic conditions (minimum and maximum ambient temperature, rainfall at site).
- Wind velocity and direction.
- Populated areas, schools, hospitals, farms, airport nearby.
- Area over which dispersion would occur.
- Ground-level concentration limits prescribed by WHO shall never be exceeded (whether during normal plant operation or during process upsets).

Chapter 8
Maintenance

Chemical plant units and machineries are subjected to dust evolved during crushing and grinding, corrosive gases escaping from charging manholes of reactors, and minor leaks from glands of shafts (agitators or pumps), from safety vents, etc. The corrosion rates are therefore generally more than the machines installed in a mechanical fabrication shop or the civil structures like sheds.

Since most of the chemical plants are run continuously, any sudden breakdown is not acceptable as it may result in unsafe condition or environmental pollution. It may also result in a long unplanned stoppage of the plant and considerable expenses for repairs.

On restarting, it requires resetting of process flow rates, adjustments of heating and cooling mediums, operating levels, etc., for the process units. Hence, it is necessary to ensure smooth run of the plant equipments and minimise breakdowns.

Managements shall carefully look into the following for proper maintenance of the process units and equipments.

8.1 Condition Monitoring of Units and Machines

Process reactors, equipments, machineries, pipelines, and other units are designed to operate at certain conditions of pressure, temperatures, flow rates, concentrations of reactants, speed of rotation of agitators, etc. The mechanical design and selection of the materials of construction (MOC) are done after considering the maximum (upper) limits of the operating conditions for safe running. Safety margins and suitable corrosion allowances are also added. This is for increasing the useful life of the equipment.

Plant-operating personnel shall regularly check the conditions at which the equipments are actually operated and the actual load on various drive motors for the process units and machinery. The observed values shall be always compared with maximum limits permitted with design values/specified by manufacturer.

However, a programme shall be in place to regularly monitor the physical operating conditions of the units. If any high noise levels, abnormal surface temperatures, or strange smells are noticed, their source shall be traced immediately and

K.R. Golwalkar, *Production Management of Chemical Industries*,
DOI 10.1007/978-3-319-28253-4_8

the reasons investigated. Vibrations in agitated vessels and any loosening of foundation bolts (of mechanical equipments) shall be examined regularly.

Thermograph cameras shall be used to check surface temperatures of furnaces, hot ducts, steam-heated units, electrical contacts, etc. High temperatures indicate some damage to internal refractory of the furnace and damage to thermal insulation and loose electrical contacts.

Senior plant engineers shall make an exhaustive list of more such observations to be made for their plant. Plant operators shall bring to the notice of senior engineers any abnormal conditions.

This will enable early detection of any likely damage of the plant equipments which may ultimately result in a breakdown. Since this can lead to an accident, environmental pollution, deterioration in product quality, or damage to plant equipment corrective action shall be taken at the earliest. *The process units may be operated at reduced load (and examined thoroughly from view glasses) if it is not possible to rectify the cause immediately. This is to check any false alarm or any undue panic.*

The plant should be stopped altogether if there is possibility of an accident and investigations should be done immediately.

8.2 Preventive Breakdown Maintenance Programme

Some typical checkpoints are given below. More checkpoints can be added as per the need for the particular plant in consultation with senior maintenance engineers and experienced workmen.

- Lubrication/greasing of bearings and rollers to be done at regular intervals (or as per manufacturer's recommendation) specially for blowers, ID Fans, and belt conveyors;
- Check the electrical trip settings for overload and correct them if necessary (hoists);
- Check the high-pressure alarms and safety valves for high pressure for critical machines (air compressors and receivers);
- Low-level alarms for boiler feed water tanks;
- Check foundation bolts, lock nuts, cotter pins, and coupling halves of high-speed/heavy-duty equipments which can vibrate;
- Junction boxes of motors, electrical contacts, and trip settings.

8.3 Facilities to Be Available in the Plant

It is advisable to have instruments and devices for detecting conditions detrimental to equipment life and for carrying out preventive maintenance. Some of these are as follows:

- Ultrasonic thickness testers for checking shells of storage tanks and process units;
- Vibration analysers for blowers, compressors, crushers and agitators or if abnormal vibrations and abnormal rubbing noise are noticed;
- Dial gauges for checking alignments of machines (specially high speed) coupled to gearboxes and motors;
- Tong testers for checking actual current drawn by a motor if ammeter is not provided or if the motor is tripping often;
- Hand-held (portable) infrared temperature detectors for detecting heat loss from furnace shell or gas ducts, temperatures of bearings (can indicate need for immediate cleaning and lubrication), and electrical bus bars (can indicate loose contacts);
- Thermographic imaging cameras;
- Dial thermometers, standardised temperature measuring instruments;
- Gas analysers and detectors for detecting malfunctions of scrubbers/process reactors.

8.4 Inventory of Mechanical/Electrical/Instrument Spares

A study of the record of spares replaced in last few months/year shall be done to know which ones are being procured more frequently. The reasons for the same shall be investigated. *Formats have been given for keeping such records which will be required for this study in Chap.* 16.

Some of the equipments might have been operated beyond their maximum limits of temperatures (heat exchangers), levels (agitated reactors), and delivery pressures (pumps).

Some preventive steps which could be taken in consultation with operating personnel are as follows:

- *The heat exchangers shall be operated at lesser temperatures by reducing the feeding rates of the reactants, or they may be provided with special quality tubes.*
- *Either lower level shall be maintained in the reactors or the agitators shall be run at lesser speeds.*
- *Equipments and piping on the delivery side of the pumps shall be redesigned and provided with pressure relief valves, as well as tripping arrangement for the drive motor.*
- *Typically, rate of charging raw material bags into the reactor can be controlled. Instead of adding all the bags together (which can suddenly overload the agitator, its gearbox, and the motor), the bags may be slowly put in one by one.*
- ***Case study***—*Instead of adding concentrated acid through a 50-mm-diameter pipe from its open end, a long pipe with a number of small holes was used. This*

enabled the addition of concentrated acid at many points in the cross section in the reactor, and it was getting mixed properly in the reaction medium. This minimised chances of pockets of concentrated acid near the shaft or agitator blades—which reduced the corrosion due to attack by concentrated acid taking place earlier.

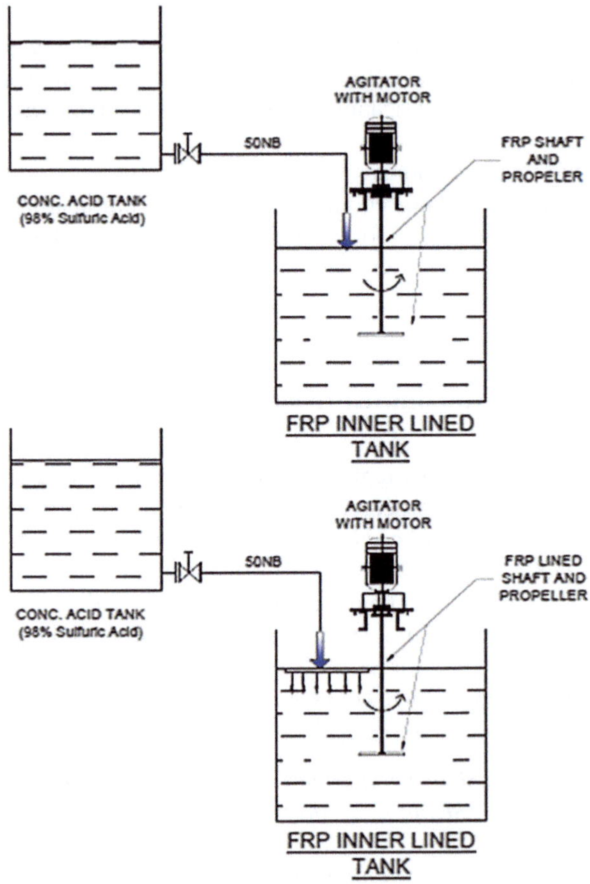

Generally, as a better practice, spares shall be procured from original equipment manufacturer only. Old spares may be repaired and reconditioned only if original spares are not available, and it is urgently required to restart the machines. But this may not be reliable and can result in sudden breakdowns.

However, some simple spare parts may be developed in-house, e.g. anti-vibration pads, level indicators, shaft sleeves for pumps, thermowells for instruments, and strainer in pipelines at suction of pumps or instrument probes.

8.5 Standardisation of Spares

It is generally difficult to maintain a large variety of spares, because the procurement needs frequent repetition of activities such as estimating future requirements, floating inquiries, obtaining samples of spares, quotations, negotiations with vendors, releasing purchase orders, physical testing and verifications of the supplies received, and their safe storage. Apart from this, there could be chances of key machinery becoming idle if some spare part is not available and interruption of production as a result. Hence, efforts shall be done by all concern to minimise the variety of spares. Management shall call regular meetings of senior chemical, mechanical, electrical, and instrument engineers to discuss these issues.

- Check whether same spares are given different names, e.g. bypass valve/flow throttle valve/water line control valve—though all are identical, but used in different sections of the plant. This can reduce the inventory of such common items.
- Is it possible to use same size of spares for different machines/process units?
- Coupling halves, foundation bolts, motors, and covers—for different pumps.
- Level indicator tubes, view glasses, thermowells, gaskets, flanges, etc., for different equipments such as reactors, receivers, mixers, and dissolvers.

8.6 Advancing the Plant Shutdown

Generally, considerable advance planning is necessary for carrying out jobs during a plant shutdown. Many spares have to be procured, extra manpower is to be arranged, some outside parties are to be engaged, and certain heavy-duty machines are to be hired (on a per hour basis at high rates usually).

However, the plant may have to be shutdown a few days earlier in certain circumstances:

- The raw materials as per required specifications are not available or are available at very high costs from a distant supplier only (not acceptable due to further addition of high cost of transport also);
- There is water shortage;
- It has become dangerous to run the equipments due their bad condition;
- There is not enough demand for the product at present due to some reason—shutdown at customer's works (or cheaper or better products have become available in the market);
- Special containers for safe packing like strong drums of required specifications are not available;
- Transport bottleneck is preventing dispatches, and storage tanks are getting filled up;

- Power supply has broken down or has become erratic due to any problem in the supply grid;
- Accidental deactivation of catalyst or damage to a key equipment (leak of boiler tubes) is creating environmental pollution or a heavy drop in process efficiency.

8.7 Postponing the Plant Shutdown

- It may become necessary to postpone the planned shutdown of the plant sometimes.
- There is good continued demand for the products (*and short supply in the market*).
- An urgent commitment is to be met for supply of products.
- Ambient conditions are favourable for better plant performance (better condensation of volatile products can take place in winter; drying of products is better in summer).
- Some essential items (like spare parts or new equipment which can be fitted only during a plant stoppage) will be available only after a few days/weeks.
- Support structures required for new equipments are not yet ready.
- Heavy-duty cranes or additional manpower (required during shutdown) is engaged elsewhere and will be available only after some more days.
- Certain modifications in process or operating conditions (e.g. catalyst operating temperatures, absorption system liquor flow/temperature) have been discussed among the operating personnel and also approved by the management. The result of these modifications/changes proposed for improvement in plant output, product quality, and/or energy efficiency can be observed only when the plant is running.

It may also have been planned to purchase new equipments for the improvement in plant. However, it will be a much cheaper option to try the modifications instead of purchasing new equipments.

In such situations, the plant shutdown for annual overhaul may be postponed by a reasonable period provided it will not violate any statutory regulation (e.g. inspections of boiler/pressure vessel/electrical installation).

Permission must be taken from statutory authorities for running these equipments for a few more days beyond the due date. All safety precautions shall be thoroughly checked before requesting for such permissions.

8.8 Cost of Shutdown

- A loss of profit can result since there will be no production.
  This can be taken care of by keeping sales going on even during stoppage of plant by building up stocks of the products before shutdown and supplying from the stored material provided there is no deterioration during storage.

To maintain good relations with customers, get the products outsourced, i.e. manufactured on contract by supplying raw materials and paying conversion charges to another industry for manufacturing the products—strictly as per specifications.

- Cogeneration products such as power, steam, hot air, or hot water cannot be sold during the plant shutdown.
- Consumption of power, water, steam, etc., may still continue to prevent solidification of products and choking of pipelines or to maintain low temperatures in storages for volatile liquids, plant lighting, water supply, operation of effluent treatment facilities, etc.
- Costs incurred due to replacement of damaged parts/repairs to machinery by purchase of new spares.
- Fuel, power, water, lubricants, and consumables for welding and gas cutting.
- Electrical, instrumentation spares and civil repair jobs (flooring, support beams and columns, and storm water drains lining).
- De-scalants for removal of scales from boilers and heat exchangers, reactor jackets, neutralising chemicals for the wastes drained out from various process units, additional labour employed for dismantling, and taking out the internal fittings from absorption/distillation towers, cleaning process equipments—
- Salaries will have to be given to personnel employed in stores, security, laboratory, office staff, canteen, and other such persons not directly engaged for production work.

8.9 Restarting the Plant After Annual Shutdown

Make sure that all statutory requirements and instructions given by inspecting authorities have been complied with. Some of these could be for boilers, economiser, steam lines, and electrical mains which have to be necessarily stopped for thorough cleaning and internal inspections.

The date of restarting the plant shall be decided after discussing with sales department. The production engineers have to produce marketable grade material on restart. However, this may not be immediately possible, and hence, the date for restart may have to be advanced by a few days (*the time it would take to stabilise the plant and to produce marketable grade material*).

Storage tanks shall be made available to store the off-spec products or arrangements made to recycle/dispose of such material till proper quality products are getting produced.

The schedule for restart shall be prepared by working the dates backwards. This will indicate which activity is lagging behind and needs to be accelerated.

Check units at the end of production chain first (product purification/filtration/distillation), cooling systems, other process units thereafter, and finally the feed end. Boxing up shall be done in the meanwhile, and start-up activities (*see examples below*) shall commence only when all units are found to be in working condition.

Example

- Heat the plant units like converter to ignition temperature of catalysts before feeding of raw materials **or**
- Start refrigeration plant for chilled water supply to condensers before starting distillation unit.

8.10 Case Study: An Annual Shutdown of a Sulphuric Acid–Oleum Plant

This is planned in the following manner:

Stock of acid is built up in storage tanks prior to the shutdown. Marketing (Sales) department is informed about the dates when the plant will be stopped and restarted. Engineering stores, human resource planning, and finance departments are also similarly informed.

The stopping date indicates when feeding of sulphur is stopped to the burner, and the starting date refers to the date when the plant operations stabilise and normal production rate is reached.

It generally takes 24–48 h (depending on the ambient conditions) to cool down and start cleaning the units after stopping the plant. It can take about 48–72 h to heat up the plant units after the cleaning jobs are over (and catalyst, acid/oleum tower internals are refilled) before sulphur feeding is started again. Thus, the plant can be idle for about four to five days in addition to the time required for carrying out the actual annual overhaul.

It is noticed many times that the production rate slowly reduces as the annual shutdown approaches due to building up of scales and dust in boilers, heat exchangers, and catalyst passes and deposits on the demisters in the acid towers. This can cause increased pressure drop in the process units, and the air blower has to deliver against considerably more *total back pressure*. Plant operators may have to reduce the airflow to the plant in such situation.

The plant is run with as much airflow and sulphur feed as possible just a few days before the shutdown. All temperatures, SO_2 % at inlet to converter and in exit gases, are checked. Airflow rate, pressure drop in each unit on gas side, and load on each motor are also checked. Any abnormality is noted down, and planning is started to correct it.

This indicates condition of each of the process unit (activity and choking of catalyst, efficiency of heat exchanger, choking of acid towers, etc.). These equipments and process units need to be thoroughly cleaned and checked during the overhaul.

A list of jobs to be done during the shutdown is prepared. This generally includes the following:

(i) Those which are necessary for Statutory inspection.

Equipments operated under pressure such as boiler, economiser, superheater, and pressurised air receiver are to be thoroughly cleaned and presented to the inspecting authority after removing insulations and cladding. Instructions for any repairs, changing of tubes, etc., are to be complied with. Setting of safety valves will be done by the inspector initially and then again when steaming is started.

Similarly, all electrical installations (transformers, metering panel, incoming high tension supply, etc.) are to be thoroughly cleaned, overhauled, and offered for inspection. The instructions from electrical inspector shall also be complied with, and all facilities tested.

Permission to continue operations is to be obtained now.

(ii) Routine cleaning of all equipments and the process units.

Sulphur melters and burner, water treatment plant internal, heat exchangers between converter passes (de-scaling of tubes), plate heat exchangers for acid and oleum cooling. Special emphasis is laid on those units having abnormal pressure drops and less efficiency (converter, acid towers, heat exchangers, etc.).

(iii) Overhauling of all pumps (for sulphur filtration and feeding to burner, acid circulation, cooling water circulation, and boiler feed water), air blower, acid transfer, and loading pumps.
(iv) Special jobs which are necessary for increasing efficiency of plant or for expansion, or diversification to more products—*which can be done during the shutdown only, e.g. replacement of the pad-type demister in drying tower by candle-type demister; loading additional catalyst in the last pass of converter; and providing an oleum boiler to generate SO_3 vapours for manufacture of 65 % oleum.*
(v) Certain repairs which are found necessary on opening process units—for example, when the catalyst was taken out for screening, it was noticed that one of the catalyst support (loading) grid was damaged and catalyst had fallen through. This should be replaced by a new one or repaired. These are absolutely necessary and cannot be done if the plant is started.
(vi) Some special major repairs—*case study of work done by the author.*

It was seen that a continuous stream of sulphuric acid was coming out from the bottom side of the drying tower when the plant was running. It was clear that even though there were three layers of acid-resistant bricks at the bottom, the seepage of acid is taking place due to damaged lining since the plant was about 12 years old. The plant shutdown was planned for **only 15 days,** and it was not possible to repair the leak by taking out the bottom layers within this time. It would have taken at least a month to take out all internals from the tower, remove the old layers of bricks, and lay new layers and cure them, before filling again with internal fittings and packings.

An enclosure of carbon steel was therefore prepared, while the plant was running and two layers of acid-resistant bricks were laid in it. The tower was emptied out on stopping the plant and lifted up slightly by special jacks. The enclosure was provided with two layers of acid-resistant bricks and acid-resistant cement paste mixed with silica and pushed below the tower. The tower was now lowered into the enclosure. The cement paste got injected inside the leaking portion (which was never seen) due to the heavy tower and got cured by itself due to presence of acid there. This tower functioned normally for more than ten years thereafter.

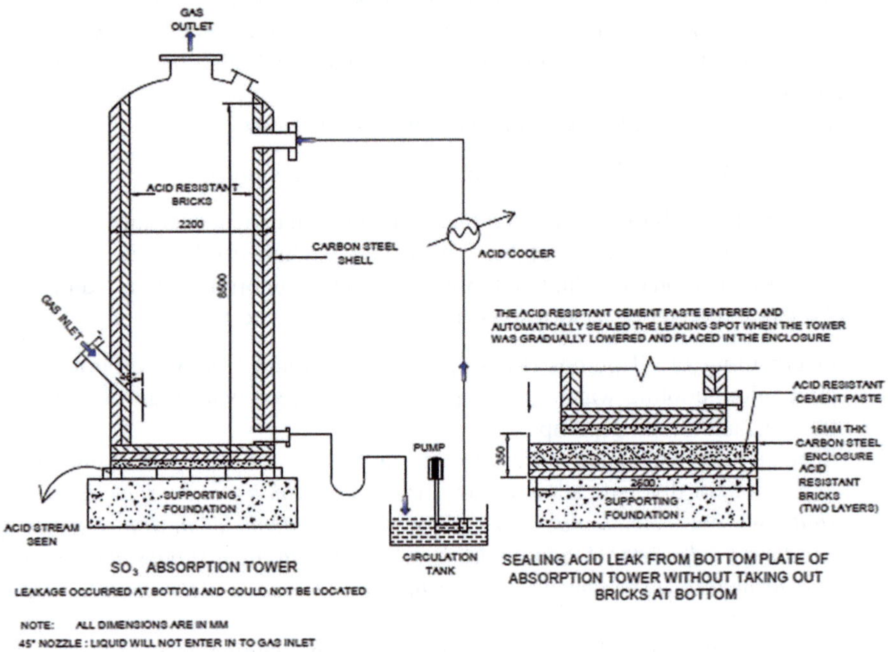

Chapter 9
Utilities

Utilities are required for operation of the main process reactors and related units for pretreatment of raw materials and processing of the products, for safe and efficient operation of plant, operation of control valves, flushing of equipments, treatment of effluents, etc. Main utilities are steam (high and low pressure), electrical power, water (for process, boiler feed, and cooling), heating mediums, inert gas (nitrogen), and compressed (and dry) air supply.

9.1 Waste Heat Recovery Boilers (WHRBs)

Waste HRBs are an important type of unit for recovery of large amounts of heat from process gases at high temperatures. It is necessary in many plants to cool these gases as per process requirements. The heat transfer coefficients are high in the WHRBs as compared to air preheaters or recuperators (which need a large area for cooling the gases due to low heat transfer coefficients). The heat is thus conveniently recovered as steam which could be at a high pressure for subsequent use for power generation and process heating elsewhere in the premises. Considerations for floating enquiry are as follows:

- Quantity of gases to be cooled, composition of gas mixture and properties of all the gases shall be known.
- Maximum temperature at which the gases will enter the WHRBs.
- Normal working pressure at which steam is to be generated, and the saturation temperature corresponding to this pressure.
- Maximum working pressure at which the boiler may be operated—to be considered for design of the boiler.
- Temperature up to which the gases are to be cooled. (D*ew point of gas mixture is to be informed to the boiler manufacturer.) The gases cannot be cooled to the saturation temperature corresponding to the working pressure of steam since some reasonable temperature difference driving force is necessary for heat transfer. Also, the gases shall not be cooled to the dew point of the gases to prevent condensation of acidic condensate on the heat transfer surfaces.*

K.R. Golwalkar, *Production Management of Chemical Industries*,
DOI 10.1007/978-3-319-28253-4_9

- Demand pattern of steam in the plant—whether it is steady or intermittently higher quantity of steam supply is required in the plant.
- Type of boiler required—water-tube type or smoke-tube (fire tube) type?
In case of fire-tube boiler, they may cost more for the same steam generation capacity since the tubes and the shell both are subjected to steam pressure. Water-tube boilers are generally used for power generation. They can start steam generation earlier than fire-tube boilers due to less volume of water in side. Only the tubes operate under pressure. The gases at low-pressure pass through the outer shell. The shell is therefore not subjected to steam pressure. Hence, cost could be less.
- Maximum pressure drop allowed on gas side. *A single pass on gas side will have less pressure drop, but may need an external bypass duct (by purchaser) to prevent excessive cooling of the gases.*
- Quality of feed water required by boiler manufacturer.
- Temperature at which it will be available in the plant.
- Vendor to inform—the orientation of all nozzles for entry and exit of gases, entry of water and exit of steam; location and sizes of cleaning manholes; overall dimensions of the boiler (length × breadth × height); space recommended all around the boiler for operation, cleaning, and maintenance.
- Vendor to supply the civil foundation drawing for the boiler full of water with all fittings and water filled in and purchaser to make civil foundations for the maximum weight. Location of foundation bolts and sliding (roller) supports shall be as per drawing given by boiler manufacturer.
- Structural saddle supports and work platforms may be obtained from the vendor.
- Vendor shall supply all standard fittings and valves for feed water, blowdown, vent, safety, as well as steam stop valves, instrumentation, level controllers with low-level alarms, and boiler feed water pumps with motors. All of these shall be supplied with test certificates and approval of statutory authorities.
- All high-pressure feed water piping along with control valves to fill up the boiler shall be designed, fabricated, and erected by vendor.
- External insulation and cladding may have to be done by purchaser after statutory inspection is carried out. **It shall be the responsibility of the vendor to get statutory approval for the entire set-up (boiler and accessories)**.
- **Erection and commissioning of the boiler shall also be the responsibility of the vendor**.
- A coal-, oil-, or a gas-fired boiler will not be a WHRB. It will need separate air pollution control systems such as dust separators, electrostatic precipitators, induced draught fans, and a tall chimney for the exit gases from the boiler.
- **Comparison of water-tube and smoke-tube boilers**.

9.1.1 Water-Tube Boiler

- Used generally for power generation and hence operated at higher pressure.
- Quicker start due to low water volume in the tubes.
- If the operating pressure is high, the saturation temperature is also high. Due to this, the temperature of exit gases is also high (which is necessary to maintain a reasonable temperature difference driving force for heat transfer from hot gases to boiling water in the boiler).
- Chances of tube leak are more.
- Cost of boiler shell is less, because shell is not subjected to high pressure.
- Ferrules are not required to be fitted in the tubes.
- Higher steam demand can be met in shorter time.

9.1.2 Smoke-Tube Boiler

- These were used in locomotive steam engines.
- It takes longer time to start due to high water volume in the boiler.
- Cost of shell is more because shell is subjected to high pressure.
- Ferrules are fitted at the inlet of tube and tube sheets are covered by refractory lining. These are required for protection of the tube inlets as well as the tube sheets.
- Higher steam demands cannot be met in shorter time.

9.2 Steaming Economisers

Flue gases exit from boilers at a temperature higher than saturation temperature of the steam *corresponding to the working pressure of steam.* Economisers are used to further recover heat from these hot flue/process gases for preheating feed water before entering the boiler.

Conventional Economisers The heat is recovered as "sensible heat" of the water, and hence, large quantities of water is passed through and comes out as hot water. System efficiency is optimum when the existing boiler(s) are able to evaporate all this hot water being fed into it. Any excess hot water will have to be used in the plant elsewhere, or it will get wasted. An alternative is to bypass the economiser from gas side. This may increase the temperatures in the downstream process equipments and affect the operations, and hence, this option may not be advisable.

It will be seen that passing a large quantity of water through the economiser coils needs considerable power to run the feed water pumps.

In case of a steaming economiser, the water which has become hot is flashed in a separate drum at a height as steam and balance water comes down to the heating coils of the unit for getting heated up once again. *The steam can be directly put in the main header or used for process heating at a pressure at which it is generated or after reducing the pressure as per need.*

Water circulation in the heat transfer coils takes place by natural convection currents as it gets heated up due to the heat given by the hot gases. No pump is required for this purpose and hence the power consumption is very low.

The feed water pump is run intermittently only to maintain the water level in the flash drum (which is a separate unit), corresponding to the amount of water evaporated. Such steaming-type economisers can be seen in a sulphuric acid plant.

Case study

An earlier design of the plant had two boilers (first after the furnace for burning sulphur and the second one after the first pass of converter) and an economiser after last pass of the converter. This economiser generated more hot water than the boilers could evaporate—especially when they were sometimes bypassed to adjust process temperatures.

The existing economiser was therefore replaced by a steaming-type economiser. Replacing with the modified economiser saved considerable power and also recovered more heat *directly as steam*. No hot water was also wasted.

A much smaller water feed pump was provided to maintain the level in the flash drum.

9.2.1 Selection Criteria

- Flow rate of gases, composition of gas mixture, gas exit temperature desired, and properties of the mixture—mean Cp, viscosity, density, etc.
- Allowable pressure drop through the unit on the gas side.
- Pressure at which steam is to be generated and the corresponding saturation temperature. This will be the maximum temperature for the hot water.
- Temperature at which feed water will be available.
- Orientation of gas inlet and outlet nozzles shall suit existing plant ducts if possible.
- Total space requirement (length × breadth × height).
- Total weight when full.
- Vendor shall supply all standard fittings and valves for feed water, blowdown, vent, safety, as well as steam stop valves, instrumentation, level controllers and low-level alarms, and boiler feed water pumps with **motors** (**optional**). All of these shall be supplied with test certificates and approval of statutory authorities.
- Feed water piping along with bypass pipe to fill up the flash drum directly shall be designed, fabricated, and erected by vendor.

- Purchaser to make civil foundations for the maximum weight. Location of foundation bolts and sliding supports shall be as per drawing given by boiler manufacturer.
- Structural saddle supports and work platforms may be obtained from the vendor and included in the purchase order.
- It shall be possible to increase the heat transfer area in future by adding more elements (more tubes).
- The tubes shall have gills of suitable corrosion-resistant material of construction for protection against corrosive gases.
- Easy accessibility of bends for maintenance (to be located outside the gas enclosure).
- Gas inlet and exit boxes, and internal fire-/acid-resistant linings.
- Refractory-/acid-resistant lining in gas inlet and outlet boxes shall be carried out by vendor to suit drain nozzles on the gas outlet boxes for draining out any (acidic) condensate.

9.3 Cogeneration of Power

Boilers are operated in many industries for generating steam required for process heating of reactors, evaporators, and distillation columns which may need steam at a pressure of 4–6 kg/cm^2 only. However, the boilers are generally operated at 10–15 kg/cm^2 or even higher pressures. Steam is also available from waste heat recovery boilers in many chemical industries at such pressures.

The pressure is brought down by means of pressure reducing valves for heating the process units. This is a waste of energy. Cogeneration is to be considered for such situations for higher system efficiency by using the exhaust steam for power generation as well as process heating, for drying, in multiple effect evaporators, for producing potable water, waste water purification, effluent quantity reduction by evaporation, etc.

9.3.1 Planning for Cogeneration

Carry out the following study for the present operations as well as future needs.

- Measure the actual steam availability: its quantity in kg/h, pressure in kg/cm^2, and temperature in °C.
- Estimate the requirement/measure the actual consumption of steam for process heating in the plant for various units—quantity in kg/h and pressure in kg/cm^2.
- Estimate the requirement of power for various units, whether they will be run together or in a staggered manner and the running hours of each.

- Similarly, estimate the requirement of steam for heating of various units, whether they will be heated together or in a staggered manner and the operating hours of each.
- This will give an idea of the peak and average requirement of power as well as that of steam required in the industry.

9.3.2 Options Which Shall Be Considered for Buying a Steam Turbine

- Backpressure-type turbine—The steam at higher pressure can be first passed through this type of steam turbine for power generation or for driving a blower/pump, and the low-pressure exit steam can be used for process heating. This is ideal for steady-state operation of the plant.
- Extraction-type turbine—It may be used if different process units require steam at different pressures. But there could be operational difficulties if too many extractions are taken at different pressures.
- Reheat turbines: The medium-pressure saturated steam from exit of a high-pressure steam turbine can be reheated by another source of hot process gases if available in the plant. More power can be generated by using this reheated steam through another turbine.

Condensing-type turbine—This is used only for power generation. All (or maximum quantity available of) the steam is passed through this turbine and the exhaust steam is fully condensed to maintain as low a pressure at the exit as possible. This maximises the power generated. The exhaust steam from this type of turbine is at very low pressure and temperature, hence cannot be used for process heating.

9.3.2.1 Examples for Cogeneration

These examples illustrate the idea that more energy can be available by using superheated steam for power generation (*by installing a back-pressure turbine instead of a fully condensing type*) and using the exhaust steam from the turbine for process heating in a chemical plant. The four cases are given for illustrating this by presenting numerical examples here.

Following properties of steam and condensate are considered for these four cases.

- Enthalpy of steam at pressure 28 Bar and temp 300 °C = 3001.3 kJ/kg;
- Enthalpy of steam at pressure 28 Bar and temp 400 °C = 3235.8 kJ/kg;
- Enthalpy of steam at pressure 4 Bar and temp 143.6 °C = 2737.6 kJ/kg;
- Latent heat at 4 Bar, and temp 143.6 °C = 2132.9 kJ/kg;

- Enthalpy of condensate at pressure 4 Bar and temp 143.6 °C = 604.7 kJ/kg;
- Enthalpy of steam at pressure 0.1 Bar and temp 45.8 °C = 2584.7 kJ/kg;
- Latent heat of condensate at pressure 0.1 Bar and temp 45.8 °C = 2392.9 kJ/kg.

Case 1 Condensing turbine, steam superheated to 300 °C

Steam condition at inlet of turbine 28 Bar and 300 °C, enthalpy = 3001.3 kJ/kg;
Steam condition at exit of turbine 0.1 Bar and 45.8 °C, enthalpy = 2584.7 kJ/kg;
Energy available for power generation = 3001.3 kJ/kg—2584.7 kJ/kg = 416.6 kJ/kg;
Energy availability efficiency = 416.6 kJ/kg/3001.3 kJ/kg = **13.88** %.
The steam from the exit of the turbine is condensed by the cooling water, and the latent heat energy is thus not available.

Case 2 Condensing turbine, steam superheated to 400 °C

Steam condition at inlet of turbine 28 Bar and 400 °C, enthalpy = 3235.8 kJ/kg;
Steam condition at exit of turbine 0.1 Bar and 45.8 °C, enthalpy = 2584.7 kJ/kg;
Energy available for power generation = 3235.8 kJ/kg—2584.7 kJ/kg = 651.1 kJ/kg;
Energy availability efficiency = 651.1 kJ/kg/3235.8 kJ/kg = **20.12** %.

Case 3 Back-pressure turbine, steam superheated to 300 °C

Steam condition at inlet of turbine 28 Bar and 300 °C, enthalpy = 3001.3 kJ/kg;
Steam condition at exit of turbine 4.0 Bar and 143.6 °C, enthalpy = 2737.6 kJ/kg;
Energy available for power generation = 3001.3 kJ/kg—2737.6 kJ/kg = 263.7 kJ/kg;
Latent heat at 4 Bar and temp 143.6 °C = energy available for process heating = 2132.9 kJ/kg;
Total energy available = 263.7 kJ/kg + 2132.9 kJ/kg = 2396.6 kJ/kg;
Energy availability efficiency = 2396.6 kJ/kg/3001.3 kJ/kg = **79.85** %.

Case 4 Back-pressure turbine, steam superheated to 400 °C

Steam condition at inlet of turbine 28 Bar and 400 °C, enthalpy = 3235.8 kJ/kg;
Steam condition at exit of turbine 4.0 Bar and 143.6 °C, enthalpy = 2737.6 kJ/kg;
Energy available for power generation = 3235.8 kJ/kg—2737.6 kJ/kg = 498.2 kJ/kg;
Latent heat at 4 Bar and temp 143.6 °C = energy available for process heating = 2132.9 kJ/kg;
Total energy available = 498.2 kJ/kg + 2132.9 kJ/kg = 2631.1 kJ/kg;
Energy availability efficiency = 2631.1 kJ/kg/3235.8 kJ/kg = **81.31** %.

Thus, overall energy availability efficiency can improve since more energy can become available when steam is superheated to higher temperature and exit steam from the back-pressure turbine is used for process heating.

The plant engineers shall choose the operating conditions (*for the back-pressure turbine*) after assessing the requirement of power and heat for their plant.

In case of a lot of mismatch between steam generation, power needed, and process heating requirements, efficient operation of the system is difficult with a single type of turbine.

A combination of small capacity turbines (extraction and backpressure type) may be considered with the balance amount steam sent to a condensing-type turbine for power generation. Cost of steam and power shall also be taken into account before the final decision is taken.

9.3.3 Points to Be Included in the enquiries for Steam Turbines

- Availability of steam—quantity, pressure, and temperature in the system. *Superheated steam is preferable—there will be more power generation with less chances of condensation inside the turbine.*
- Steam consumption of the turbine in kg/KWH (*steam rate*)—**by vendor**.
- Speed governor and steam flow controller arrangement—**by vendor**.
- Speed of turbine, gear box, and generator (RPM) during operation. The steam turbine, gearbox, and generator shall be supplied as preassembled unit.
- Lubricating oil system—cooling arrangement, oil tank, oil pumps (directly turbine shaft driven or through separate motor) with a standby pump and should auto-start in case one pump trips; and safety valves at the lube oil pump discharge (should be duly tested again at purchaser's site).
- Battery-operated emergency lubrication pump, battery, and charger.
- Material of construction of all the critical parts, e.g., steam nozzles, turbine blades, shaft shall be informed by the vendor. They shall be suitable for the maximum temperature of the superheated steam.
- **Scope of supply** to include speed governor, hand controls, emergency trips, moisture separator with steam trap for incoming steam and instrumentation (as below).
- Instrumentation: pressure gauges for steam and lubricating oil, bearing temperature indicator, electrical panel instruments (voltmeter, ammeter, KW, KWH, frequency meter, etc.). Tachometer with digital remote indication, running time totaliser.
- In case of a condensing turbine, a complete package (steam condenser, condensate extraction pump, vacuum pump/steam ejector (*if possible, cooling tower and water circulation pumps, valves, and piping also*)) shall be included in scope of supply.
- Overall dimensions and weight of the assembly.
- Drawings for support structures and details of civil foundation to be given ***by vendor***.

9.4 Steam Superheaters

Superheating of steam can increase the power output of the turbine and minimises the condensation of steam on the blades. The superheater can be used to recover heat from hot gases. Considerations for the selection of the superheater.

- Pressure, temperature, and rate of flow of steam.
- Pressure drop through the superheater shall be minimum on steam side as well as gas side.
- Temperature of the steam required at the outlet of the superheater.
- Nozzles for steam inlet and outlet.
- Provision of valves for steam inlet and outlet, safety, fitting of pressure gauges, vents, drain points for any condensate formed on gas side.
- Possibility of increasing the heat transfer area by addition of a few more tubes in future.
- **Hot Process Gas** (if available in plant): Composition and flow rate, temperature of gases at inlet and outlet, presence of any corrosive component in the gas mixture and the dew point (the temperature at which the corrosive component might condense and *reduce the life of the superheater tubes if the gases get cooled to this temperature)* or requirement of auxiliary fuel for superheating.
- Provision of C.I. gills on the tubes for protection against corrosion.
- Size and orientations of gas inlet and outlet nozzles—(better if they suit gas ducts at site) internal lining of refractory/AR bricks in the enclosure for gases.
- Purchaser to make civil foundations suitable for the maximum weight and the overall dimensions (length, breadth, and height). Location of foundation bolts and any other supports shall be as per drawing given by vendor.
- Structural saddle supports and work platforms may be obtained from the vendor.
- Vendor shall obtain approval from statutory authority for the unit as well as all the valves and mountings since this is covered under Boiler and Pressure Vessel Regulations in all countries.
- Vendor shall clearly state the battery limits and exclusions from their scope of supply.

9.5 Steam De-Superheaters

Steam can get overheated by the superheater if the flow is less due to any reason. This can be harmful to the turbine blades if the temperature of steam is too high. The de-superheaters are used to keep the steam temperature in control.

The temperature of the superheated steam can also be controlled to the desired temperature by bypassing the hot gases around the superheater.

The de-superheater has specially designed nozzles for spraying a controlled amount of treated water in the superheated steam. An enquiry for steam de-superheater shall include the following:

- Conditions at inlet: steam pressure, temperature, flow rate: in kg/h.
- Conditions desired at the outlet: steam pressure and temperature. The temperature should not be too low to prevent condensation of steam till it reaches the steam turbine inlet.

- Material of construction of de-superheater body and water injection nozzles.
- Provision of automatic control valve for feed water injection.
- Provision of external reinforcements for the shell (it may collapse due to spray of excess water when steam supply is too less.)
- Orientation of all incoming and outgoing nozzles for steam (since all steam and water piping involved are covered by statutory regulations). Alternatively, all such connecting high-pressure piping shall be carried out by the vendor.
- Instrumentation: to include automatic water flow control valve for spray injection.
- Moisture entrainment separator to be supplied by vendor.
- Vendor shall clearly state the battery limits and exclusions from their scope of supply.

9.6 Electrical Equipments and Installations

Electrical power is one of the most important inputs for a chemical industry. Requirement of power, type of installations for drawing power from external grid, using it safely and efficiently in the process units and machinery should be looked into carefully by the plant engineers as well as senior management.

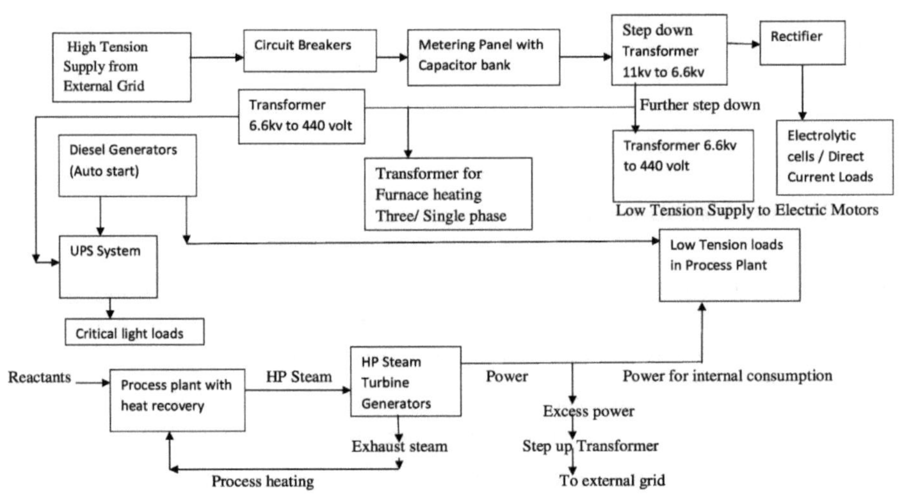

A typical electrical installation in chemical industry

9.6.1 Electrical Power Needed in Initial Phases—Before Start of Plant

Estimate the requirement of power during construction, erection, and trial runs of various process units and machineries. This can be done by knowing the horsepower of the motors connected to various loads and the expected running time for various trials.

Many welding transformers, heavy-duty cranes, flood lighting, conveyors are in use during the construction phase of the plant. These are in operation for the first few months only and gradually their use gets reduced.

The mechanical trials of machineries can be planned in a manner to avoid heavy electrical load on the main incoming supply by running them one by one if possible. The trials for all machines can thus be completed by running less machines at a time.

The **dry run** of the plant can now be taken by running almost all of the units *as they would run eventually* but without feeding the raw materials. The power required for the dry run will be generally a little less than the actual load of the plant when it is run at its rated capacity.

Hence, sanction (agreement with power supplier) for the initial power requirement (during construction and erection phase) from external grid shall be obtained for a *lesser* load according to scheduling of above activities. This can minimise the power bills which can include the fixed charges per KVA of connected load in the construction and erection stage (*which can take a few months to complete*).

However, if sanction for the power supply from external grid is obtained on the basis of the full running load of the plant, the fixed charge per KVA can be much higher than above which *will unnecessarily increase the project cost*, since the grid authority will be committing a much higher power supply from the beginning of the construction stage itself.

9.6.2 Electrical Power Need

A steady supply of power will be needed from commissioning onwards.

Types of power supply required

- Alternating current for induction motors and furnaces (three phase and single phase) and
- Direct current for electrolysis plants.

Electrical loads of the plant units and machineries shall be carefully estimated after their designs are finalised. Following matters shall be considered.

9.6.2.1 Running Load

Running load of the plant units is at rated capacity and at maximum production rates with existing machineries. This shall be rationalised on the basis of actual motor HP, the various machines, and process units which may be run *simultaneously* and the hours in a day for which each of the motor will be run. The total connected load may be reduced by staggering the running of higher HP motors.

9.6.2.2 Installed Load

Installed load, i.e., total load of all units (running machineries plus standby units)—This is generally required by power supply agencies/external grid supply authorities since the facilities for main incoming supply are to be designed accordingly. These include high-tension transformers, metering panels, supply lines, and support structures.

Special consideration is required for heavy loads such as crushers, high capacity electric furnaces, big pumps, or blowers which may cause fluctuations in the incoming supply grid and disturb other nearby industries or power consumers. Hence, arrangements must be provided in the equipments to start on a minimum load and then slowly increase the load. Some of the methods for this could be provision of variable-frequency drive motors, throttling inlet/exit valves, controlled feed rates of solid lumps to crushing machines, etc.

9.6.3 Sanctioned Maximum Demand

The significance of this shall be very clear to plant operating personnel and electrical engineers in the plant and senior management.

The total power consumption is automatically monitored at regular intervals (generally at 30 min). Any excess draw of power beyond the sanctioned maximum demand can attract heavy penalty from the grid authorities. Hence, it should be monitored by the plant personnel also and electrical loads shall be reduced in time if there is a chance of excess draw of power.

A warning device shall be installed if the actual draw of power is going to exceed the sanctioned maximum demand (to prevent such penalties).

Installation of *separate* energy meters shall be found very useful for monitoring individual power consumption of main process units, auxiliary units, utilities, lighting, offices and restrooms, effluent treatment plant, and any other equipment if felt necessary. It will become possible to monitor their consumptions and make efforts to minimise unimportant loads.

9.6.4 *Emergency Loads*

Typical units and machineries to be considered as emergency loads:

- Lighting at important units,
- Important devices for safety and control instruments for process,
- Boiler feed water pumps,
- Fire fighting pumps,
- Exhaust and scrubbing systems for air pollution control,
- Effluent treatment plant units.

More equipments can be added to this list as per need in the plant.

The sequence of starting various equipments is to be decided by plant engineers (generally safety systems, control instruments, and boiler feed water are started first and are followed by effluent control and scrubbing systems. Production units are started in the end).

9.6.4.1 Power Required for Other Equipments

Power required for other equipments which are run intermittently—such as hoists, cranes, belt and bucket conveyors, filters, and water pumps for filling up overhead tanks.

9.6.4.2 Maintenance Facilities

These are loads due to welding sets, drilling machines, fabrication machineries, travelling cranes, hoists installed in the (centralised) workshop. These facilities are very useful for in-house repairs of plant equipments and for some fabrication.

9.6.5 *In-House Power Generation*

Chemical industries need a steady continuous supply of power since any outage for even a short time can suddenly stop all running units. This can create unsafe conditions, environmental pollution, loss of production, and sometimes even jamming of agitators or choking of pipe lines. It then needs considerable time and efforts to resume proper operations, e.g. *feeding of raw materials, flow rates of heating and cooling mediums to process reactors have to be reset.*

Captive power generators may be therefore installed in many chemical plants which are to be run by steam turbines. These can get the steam supply from coal-fired boilers or from waste heat recovery boilers operating on hot process gases. They can reduce dependence on supply of power from external grid. They can also increase energy efficiency of the plants since steam is available in-house for cogeneration (and process heating as well). The external grid supply may have the following drawbacks at some locations in the country:

- High cost of power,
- Occasional voltage fluctuations may trip drive motors even when power supply is continuously available otherwise, and
- Failure of power supply for even for a few minutes can disturb plant operations.

The power generated in-house by the captive generators can be either more **or** less than the requirement for running the plant units and machinery.

9.6.5.1 Excess Power Generated

Excess power generated (more than own requirement): The additional power can be exported to external grid. This is not easy, however, due to conditions laid down by grid authorities for quantum of power and reliability of supply.

The practical issues of ensuring steady export of power (24 h per day × 7 days per week) without any fluctuations in voltage or frequency and providing a substantial amount of power, say 400 kw, to a few megawatts as required by the grid authorities at a selling rate decided by them are to be addressed. Special agreements (known as power purchase agreements) are made accordingly between external grid and plant managements.

This may be possible in case of an excess in-house power generation (from a large quantity of steam available from WHRBs and other units). Such plants can continue to run due to internal power generation even when the grid supply fails.

However, it may be very difficult to supply power to the grid in case of a *small chemical plant* (even with WHRBs units) since enough steam may not be available steadily due to variations in process conditions. As a consequence, the turbines may not be able to generate steady and sufficient power for supplying to the grid. It may not be acceptable to grid authorities who generally want a steady supply.

9.6.5.2 Power Generation Is Less Than the Requirement

The options to be considered are as follows:

1. Feed whatever power is generated to units which are not very important and continue to operate the remaining units in the plant on external grid power supply (which is used for making up the shortage). Separate feeders may be required for separate units with facility to draw power from both sources in case of need (or emergency).
2. Install an induction generator with excitation provided by the external grid. This will automatically match the phase sequence and frequency of the two supplies. The draw of power from the external grid gets reduced automatically (to the

extent of own internal generation), and thus, the amount to be paid for power consumption also comes down. This happens continuously and automatically even when the internal generation is not steady (due to unsteady steam availability because of some operational adjustments in process plant). It is thus possible to optimise use of steam.

3. (*Please refer to figure given at* 11.6.2 *in Chap.* 11 *on Energy Efficiency*)
   An equipment requiring considerable power has a motor and a steam turbine also attached to it (with *an extended shaft in some designs*). As the WHRB supplies steam to this turbine, it shares the load on the electric motor—which is always kept on—and the draw of power from grid reduces accordingly. The speed of the driven unit remains steady because the motor is *always kept on.* This arrangement can work for a sufficiently heavy load even when steam generation is **not constant** (due to occasional bypassing of the WHRBs as per process requirement).
   In case of smaller loads, more such arrangements would be required because there is no electrical generator (each smaller load will need a separate turbine and a motor). *This may need too many equipments to the plant and complicate matters for maintenance engineers.*

While floating enquiry for an electrical generator, the vendor shall be asked to give following details:

- The type of unit offered (induction/synchronous).
- Maximum continuous rating (KVA) for power generation.
- Shall be a three-phase unit. Frequency shall be 50 cycles/s or 60 cycles/s depending on the common prevalent practice in the country.
- Voltage (shall be 415 ± 3 %) and current available at terminals (shall be suitable for the maximum current rating and some overload current).
- Maximum weight of the unit.
- Normal operating speed and maximum speed permitted.
- Suitability of the design and construction of the generator for the ambient conditions at the location of the plant (purchaser shall inform the maximum and minimum ambient temperatures, humidity, rainfall, presence of any dusty conditions at place of installation). Cooling fan shall be available.
- Details of base plate and foundation required for installation of the generator and gear box assembly (if required).
- Test certificate shall be provided for resistance between windings, winding loss, and earth fault test with body of the unit.

9.6.5.3 Gas Engines

Gas engines are installed (to run electrical generators) where piped gas (from external gas grid connected to oil wells or other source of natural gas) is available. The incoming gas lines shall have pressure regulators, safety valves, and relief vent connection to flare tower and flame arrestors. Appropriate automatic drain valves shall also be provided for removal of condensates, water, etc., from the gas lines. *Material of construction of the gas pipes shall be corrosion resistant since the gas may contain moisture and traces of sulphurous gases.*

9.6.6 Uninterrupted Power Supply for Critical Loads

This shall be available for reactors producing high-value chemicals, computerised process controls, double-door sealing systems for units handling explosive materials, safety vent valves, and cooling systems for highly volatile or inflammable fluids.

This can require a bank of batteries, inverters, switch for an **immediate** automatic change over when main power supply fails due to any reason. This system can be quite costly depending on the connected electrical loads and the duration for which the assured supply is required.

9.6.7 Compliance with Statutory Rules and Instructions

For safety of transformers—special enclosures shall be made, drain pits provided below the transformers to collect oil leaks,

Provision of *special fire fighting equipment* for all electrical installations.

Earth connections for electrical installations as well as process units, storage tanks, pipelines, and gas lines handling inflammable materials shall be as per instructions/standard practice.

Electrical cables should be away from pipelines carrying acids, hot gases, and steam. There shall be no chance for entry of strong acids or alkalis in the cable trenches. The cables shall be laid at a safe distance from such pipelines.

Safety audits of all electrical installations and equipments shall be regularly carried out as per statutory rules and immediate corrective actions taken wherever necessary. Reports of such audits shall be available on demand by inspecting authorities.

9.6.8 Electrical Transformers

The product mix and the production capacities planned for the present and as per planning for the future shall be considered for making a list of the process units and utilities required. The list shall clearly indicate horsepowers (*as well as operating voltage, phases, and frequency*) and running hours of each motor which will be operated. Other types of electrical loads such as lighting, furnaces, electrolysis shall also be listed because they could be single phase load or DC loads.

Senior engineers shall estimate the present and future electrical load on the above basis. The main transformer for incoming power supply from the grid and other internal transformers shall be accordingly procured. Some of the considerations are as follows.

- *This list is indicative and senior electrical engineers shall prepare an exhaustive list of features required for procurement.*
- Total running and installed load at present and in future—in each phase (for balancing the loads in each phase).
- Incoming supply voltage and outgoing voltage desired at normal operating conditions.
- Maximum voltage/current/power KW available on secondary (LT) side in case of step-down transformer (which may be used for electrically heated furnaces) at various output voltages.
- Maximum current permissible (on HT side) on primary side at various power outputs as above.
- Whether the output voltage can be varied "on load" or the unit will have to be switched off (off load type) before changing taps? [frequent switching on/off may not be permitted by the local electrical authority if the connected load is heavy].
- Installation: whether it will be outdoor or indoor.
- Maximum temperature rise permitted above ambient temperature in summer at the location where the transformer will be installed. Whether natural or fan cooling arrangement is provided?
- Total weight after oil has been filled in.
- Properties of oil (dielectric strength of oil, chemical composition, and moisture content) and its quantity filled in. (Vendor shall submit test certificate for the oil from an accredited standard laboratory.)
- Standard fittings and mountings such as breather (with drying agent), junction box for (incoming) leads, and outgoing (cable/bus bar contacts) connection.
- Necessary test certificates for hysteresis losses at various loads, insulation tests for the windings and shell of the unit.
- Total space required all around for maintenance shall be considered.
- Any instructions from electrical inspector/SEB for installing the new transformer (especially when capacity of the process plant will be increased)—earth connections and separate earth pits for the transformer body and incoming neutral.
- **Rectifiers**—The output power (KW), DC voltage, and maximum current available should match the requirement of very high currents for electrolytic processes (e.g. electrolysis of brine to produce caustic soda). The rectifiers shall be full-wave type and electrical efficiency (DC power available per unit AC power input) should be confirmed by the supplier. Design of cell groups shall be done accordingly. Other considerations will be similar as above.

9.6.9 Electrical Power Distribution in Plant

Management shall discuss the following main points with senior chemical, electrical, and mechanical engineers and arrange the power distribution in the plant and premises accordingly.

- Total actual running and installed load at present and in future—in each phase (for balancing the loads in each phase).
- Incoming supply voltage required for various units at normal operating conditions.
- Classification of load.
  - Inductive (e.g. electrical motors) or resistive (e.g. furnaces).
  - Single phase or three phase.
- Balancing of loads in each phase.
- Any DC load installed for electrolytic processes.
  - High-torque motors—for heavy loads with high inertia.
  - Electrolytic load—for deciding ratings of rectifiers.
- Running hours and operating timing of all electrical loads. (*High-power-consuming equipments may be run only during the timings when tariff could be less—**in certain countries**.*)
  - Capacitor banks to be designed to have a power factor greater than 0.99 for efficient use of power.
- Main bus bars: Compare aluminium and copper bus bars for the intended location, total current carrying capacity and presence of corrosive gases, if any.
- Starters, push button stations, flameproof fittings, motors, and earth connections as per applicable safety standards. Approval shall be taken from electrical inspecting authorities before commissioning.

Plant Lighting The process plant must have sufficient lighting in the important operational areas where control valves, level indicators, and pressure and temperature gauges are located. Good lighting shall also be provided at all escape routes, ladders, work platforms.

Areas having inflammable vapours should have flameproof lighting. Avoid proximity of high-wattage lamps to units in such areas as the vapours can ignite due to the heat generated by such lighting.

Provide automatic switching on in case of power failure in the lighting circuit.

Provision of electric supply from separate feeders for alternate light fitting so that in case of failure of a particular feeder, there will still be some light available.

9.6.10 Electrical Motors

These shall be selected by consideration of the following criteria:

- List of all process equipments, pumps, blowers, and machineries which need a drive.
- Maximum running load at present and expected load in future when the plant capacity is expanded or new products are added. (Some process units may not be changed, but due to higher load they will have to be provided with higher HP motors.)
- List of machineries which need high starting torque (e.g. crushers and grinders, stirred tank reactors where agitators may need heavy starting power, loaded conveyer belt drives, hoists, etc.) These machines may have to be provided with gear boxes, belt drives, and variable-frequency drive motors.
- Individual requirement and frame size shall be looked into in detail.
- Electrical specifications: power supply frequency, voltage, phases (single phase or three phase) (*supply frequency is not same in all countries—it is 50 Hz or 60 Hz and the rotational speed of motor* **RPM** *depends on it*).

Areas in the plant where the motors would be installed

- Indoors or Outdoors.
- Maximum ambient temperature and humidity at place of installation.
- Presence of dust, vapours of corrosive, inflammable, or explosive chemicals (flameproof motors will be required).
- Type of speed control required (stepwise or continuously variable).
- Position of motor with respect to the equipment (vertical or horizontal).
- Junction box: leads from windings shall be accessible for connection and inspection at external contactors. Junction box cover should prevent ingress of dust, gases, and moisture).

9.6.11 Diesel Generator Set

These are generally provided to supply power for emergency loads rather than for running the entire plant.

Managements shall carefully consider the installation of a DG set where the possible consequences of any sudden power failure in a continuously operating plant could be:

- Unsafe conditions, e.g., pressure rise in reactor or runaway exothermic reactions.
- Escape of dangerous vapours if the refrigeration facility stops.
- Water level in boilers dropping very low.

- Certain equipments may get jammed due to formation of some thick slurry or pipelines may get choked due to solidification of the contents.
- To maintain lighting.
- Environmental pollution by interruption of scrubbing liquor supply to scrubbers.

Selection Criteria for DG Set

- Total individual loads to be run on DG set,
  - Process plant load,
  - Emergency load,
  - Utility services,
  - Lighting requirements, and others.

The power rating of the DG set should be enough to provide power for all essential units continuously. *Senior shop floor engineers shall decide the order of importance and starting sequence of various motors and other loads.*

Net electrical energy available at generator terminals as KWH per litres of fuel consumed (i.e. fuel efficiency).

- Are soft starters/fluid couplings/VFD provided for heavy load?—*if these are not provided, there could be sudden heavy load on the DG set and it can trip.*
- Maximum starting load the DG set can take—to match with emergency loads planned by purchaser.
- Generator: maximum current and voltage available at terminals,
- Maximum ambient temperature at the location and maximum rise permitted *above ambient* by the manufacturer.
- Engine cooling arrangement provided by manufacturer.
- Standard accessories should be available along with the DG Set: **instrumentation for e**ngine/generator tachometer, engine temperature, indication for electrical power drawn in kilowatts, voltage, ammeter, KWH meter, frequency meter, start up batteries, and battery charger.
- Fuel oil tank and strainer. (Fuel shall be always available for at least eight to ten hours of continuous run at full capacity of the diesel generator set by providing additional storage tank if necessary.)
- Noise level at full load operation shall be less than required by the statutory authorities; generally, it shall be less than 70 dB at one metre distance from the DG set.
- Exit gases from the DG set shall meet statutory norms for environmental pollution.
- Cost of electrical switch gear (isolation switches, connecting cables, metering panels, etc.) required for using the DG set shall also be considered.
- The DG sets should get automatically switched on when main power supply fails.
- Initially only one set shall start, and when more loads are being added, another set shall start automatically to share the load (when the load on first set increases to certain set point of its capacity).

- Total space required for installation for DG set and its accessories.
- Possibility of installing heat recovery steam generator/air heater from exhaust gases of large DG sets.
- Total cost of power generation to be calculated by taking into account the following:

Cost of fuel and lubricants, operators salary, annual maintenance costs, interest paid on investment in the installation (DG set, accessories, separate shed, electrical switchgear).

9.7 Refrigeration Systems

These are required to operate the process plants at low temperatures, production of organic chemicals, crystallisation and separation of crystals from concentrated solutions, condensation of volatile solvents, etc.

9.7.1 Essential Components

Compressor and condenser for the refrigerant vapour, receiver and evaporator for the refrigerant liquid, cooling water pump, chilled water/brine circulation pumps, instrumentation, cooling tower, etc. Common refrigerants are Freon and ammonia

The refrigerant vapour is compressed by the compressor and sent to a (generally water cooled) condenser. The liquid is collected in a liquid receiver from which it is fed *at a controlled rate* to the evaporator where it evaporates (by taking up heat from the circulating chilled water/brine) and the vapour is then again compressed by the compressor.

Small capacity units have air-cooled condensers.

The sensible heat of the compressed incoming refrigerant vapours as well as the latent heat of condensation is ultimately removed by cooling water flowing through the condenser. A cooling tower is used to maintain the temperature of the cooling water for proper condensation.

The refrigeration units consume considerable power and have many components operating at higher pressure—which are to be treated as pressure vessels with all applicable regulations.

9.7.2 Criteria for Selection/Procurement

- Total heat to be removed from the system by the circulating chilled water at present in summer and winter seasons. This should also be estimated when the plant capacity will be expanded in future. This heat will be taken up by the liquid refrigerant and it will evaporate. The compressor will increase pressure and temperature of the vapour and send to the condenser (through which cooling water is circulated).

- Normal and lowest temperatures required for the chilled water while circulating through the process units like condensers and cooling jackets. Maximum flow rate of chilled water shall also be considered while deciding the capacity of the pumps.
- Chilled water/brine circulation pumps, circulation tank, cold insulation on tank, and piping are important components of the system.
- An important consideration is to provide a warning lamp and a siren if the chilled water supply to process plant units reduces due to any reason because uncondensed vapours may escape—these could be toxic, or inflammable.
- It may be found more practicable to install two units of lower capacity (say 200 TR each) when total capacity of 300 TR is required rather than one single unit of 300 TR. In case of any breakdown, the process plant will continue to get refrigeration from at least one of them.
- Vendor shall be asked to provide expected power consumption per tonne of refrigeration loads at lowest temperature conditions of the chilled water/brine. *The power consumption will reduce while operating at higher chilled water temperatures.*
- An efficient compressor (oil-free type may be considered) for the refrigerant.
- Cooling tower of a little additional capacity shall be selected to take care of reduction in capacity in humid weather and fouling of condenser heat transfer surfaces.
- *It may be noted that refrigeration system operating on high-pressure ammonia is to be operated and maintained very carefully to prevent any accident.*
- Capacity control facility to run the compressor at lower load if required.
- The compressor may be provided with a variable-frequency drive motor.
- Vendor may supply screw compressor with inbuilt safety devices for tripping the motor in case of jamming/high discharge pressure due to any reason.
- Safety valves shall be provided on compressor, and condenser to prevent excess pressure in condenser due to any reason (interruption in cooling water flow at inlet of condenser, high cooling water temperature in summer)–*automatic tripping of drive motor for compressor shall also be provided when such excess pressure occurs.*

  Please refer to Fig. 6.4 in Chap. 6 Safety Management.
- Instrumentation: temperature gauges and cut-outs for chilled water/brine at inlet and outlet of evaporator; pressure gauge for gas at inlet to condenser, pressure gauges for vapour at compressor inlet and outlet; liquid receiver inlet; and lubrication oil pressure gauge.

- Level controller for flow of liquid refrigerants to evaporator to prevent high level so that liquid drops may not enter compressor.
- Initial supply of refrigerant and lubricant for commissioning to be made by vendor.

9.8 Cooling Towers

These are required to supply cold water to cooling jackets and coils of process reactors, to condensers for volatile chemicals, to condensers of refrigeration plants.

The incoming hot water evaporates partially and gets cooled.

9.8.1 Selection Criteria

- Duty conditions: To cool m^3/h of water from T_1 °C to T_2 °C.
- The plant engineers shall inform the vendor the minimum and maximum relative humidity, height above mean sea level and wet bulb temperature (design value) in different seasons of the year at the plant site. Maximum ambient temperature shall also be informed.
- Whether the cooling tower will be installed on ground floor or on roof. The latter can save some floor space, but it needs a strong roof. It can be useful as storage of cold water for emergency use in process plant.
- Type of cooling tower draught:

  - Natural draught type has more height and does not need power to operate;
  - Induced draught created by a fan can blow away the moist air more upwards —but it will need power to operate the fan (fibreglass reinforced plastic blades are preferred).

- Material of construction—It should be able to withstand ambient condition in the plant.

Chemically treated wood or fibreglass reinforced plastic with stabiliser may be used.

- Cement concrete basin (if required by the vendor) drawing to be given by vendor for the cooling tower.
- FRP basin and support structure should be supplied by the vendor, along with accessories—ladder, induced fan–gearbox–motor assembly, protective grill at top, vibration monitor for fan–gearbox–motor assembly, thermostat for switching off/on the fan during low ambient temperature, blowdown valve for draining the basin.

- Float-operated valve for make-up water to maintain the water level in the basin. *Only treated water shall be used as make-up water to minimise fouling of heat transfer surfaces in equipment like condensers, cooling coils, and jackets. Suitable chemical treatment shall be done to prevent growth of moss, algae in the water.*

9.9 Air Drying Plant

Dry compressed air is required for operating diaphragms in control valves for process streams, for flushing out residual liquids in pipe lines, for operating drain valves of certain reactors, for opening vent valves, etc.

The main components of an air drying plant are as follows which shall be checked thoroughly while procuring the plant.

- Air compressor and air filter, oil separator.
- Air cooler, or electric heater, solenoid valves, non-return valves, etc.
- Air receiver (this must be a vessel tested at 1.5 times the working pressure and must have safety valve, rupture disc, automatic moisture drain valve. It shall be inspected and permitted for use by the local statutory authorities).
- Automatic on/off switch for the air compressor as per pressures set on the air receiver.
- Instrumentation necessary: pressure and temperature gauges, electrical controls for heaters, high temperature cut-outs.

Various methods are used for removal of moisture from compressed air: (i) by absorption in silica gel and then regenerating it by electrical heating or by using heat of compression and (ii) refrigerated type for removing moisture by condensation.

9.9.1 Considerations for Procurement

- Volume of dry air required for process/instrumentation in nm^3/h.
- Dew point temperature for compressed air as specified by the instrument manufacturer.
- Pressure at which dry air is required in kg/cm^2.
- The refrigerated type is for higher volumes of dry air required. It cools the incoming air and removes the moisture by condensation.
- Refrigeration load TR for cooling the incoming air.
- Cooling medium employed (chilled water or chilled brine): temperature and flow rate of the cooling medium required and details of moisture removal arrangement.

9.10 Nitrogen Plants

Nitrogen is an inert gas and can be obtained from atmospheric air. It is used in the chemical industries for providing an inert atmosphere during testing and cleaning of pipelines and vessels for hydrocarbons or inflammable materials, in pharmaceutical industries, in the food packaging industries, for flushing of process units prior to opening for inspection. It is also used for sealing material feeding nozzles for waste incineration plants operating under suction conditions.

Criteria for procuring a nitrogen generation facility are as follows:

- Purpose for which it is required.
- Pressure at which it is required.
- Purity of nitrogen desired.
- Quantity required—small amounts can be bought in cylinders while an own generation facility can be set up for getting a continuous supply for process use. (*Handling liquid N_2 could be dangerous.*)

The two main processes used for onsite generation are as follows:

- PSA plant can generate N_2 up to 99–99.999 % (for critical applications).
- Membrane plant can generate N_2 up to 95–99 %.

9.10.1 Pressure Swing Adsorption PSA Plant

Atmospheric air is filtered, compressed, and cooled to remove dust, oil, and moisture. The dry compressed air is passed through two pressure swing adsorption (PSA) towers which operate alternately. These have:

(i) Activated alumina bed (to remove moisture and CO_2) *and*
(ii) Carbon molecular sieve bed which adsorbs the oxygen gas at higher pressure.

N_2 comes out as product and is sent to a surge tank.

If very pure N_2 is required, the remaining O_2 is removed by combining with a stream of injected hydrogen in the presence of palladium catalyst producing water vapour. The gas is cooled and dried to remove water.

Twin tower units with copper catalyst are used alternately (if H_2 is not acceptable) where in copper combines with oxygen to produce CuO, and very pure N_2 is obtained as the product after cooling the exit gas. It is stored in a receiver.

The copper catalyst is regenerated by reacting CuO with hydrogen. The gas is purged out during this operation.

Main system components to be checked before procuring the N_2 generating facility:

- Air compressor with filter at suction,
- Refrigerated air cooler,

- Chilled water supply at 8–10°C,
- Air receiver with automatic moisture drain valve,
- Pressure swing adsorption towers with activated alumina and carbon molecular sieve supported on stainless steel perforated plates,
- Surge vessel for N_2 gas (volume shall be sufficient to take care of pressure fluctuations during pressurisation and de-pressurisation),
- De-oxo unit with palladium catalyst/copper catalyst (if H_2 is not acceptable),
- H_2 injection controller (hydrogen cylinders may be procured as per need),
- Gas cooler for exit gases,
- Online oxygen analyser unit, temperature, and pressure gauges,
- Pressure regulators,
- Safety valves on all pressurised units,
- Hydrogen supply—on site generation plant.

Purity of product N_2 when H_2 is used for removing O_2:

O_2—Up to 3 ppm, H_2—0.5 %;
N_2—Balance;
Dew point depends on moisture removal efficiency;
If hydrogen in product exit gas is not acceptable, copper catalyst-based de-oxo units are used to remove the oxygen.
Purity of N_2 gas produced in this case:
O_2—1 ppm, H_2—NIL;
N_2—Balance (99.999 %);
Dew point—depends on moisture removal efficiency.

9.10.2 Membrane Process for N_2 Gas

It depends on selective permeability rates of different gases through membranes (made from hollow polymeric fibres). Gases such as H_2, NH_3, CO_2, and O_2 are able to permeate through the membranes faster than N_2.

Purity of N_2 coming out:95.0–99.00 %.
This is not suitable for critical application.
The purity required could be 99 % onwards for critical application.

To examine the following carefully:

- All pressure parts shall be as designed and fabricated per standard codes (ASME section VIII) while using only certified material of construction.
- Radiography shall be done for all pressure parts and all connecting piping between components up and the final discharge pipe of N_2 if it is taken out under pressure.
- Post-weld heat treatment shall be done for all welded parts and certified.

- All bought-out components and spares shall be certified for high-pressure service, e.g. toughened sight glasses.
- All units operating under pressure shall be provided with safety valves, e.g., air receiver, air dryer, PSA vessels, Nitrogen receiver, etc.
- It is better to have oil-free compressor. These will ensure longer life for carbon molecular sieves and improve performance of the plant.
- Gas coolers shall supplied with chilled water/brine at as low a temperature as possible for complete removal of moisture for getting better performance of the carbon molecular sieves.
- Provision of online gas purity analyser for product nitrogen gas. The oxygen and hydrogen content must be analysed for critical applications.
- Carbon molecular sieves shall be certified for selectivity for O_2, CO_2, crushing strength, surface area m^2/m^3 of packed volume, density.
- Supplier shall load in sufficient amount to reduce frequency of regeneration.

9.11 Water Budgeting and Management

A very large number of chemical industries require water for the following uses:

- Making solutions of reactants for further processing (water which combines chemically and forms the product or as water of crystallisation), dilution of process streams if required;

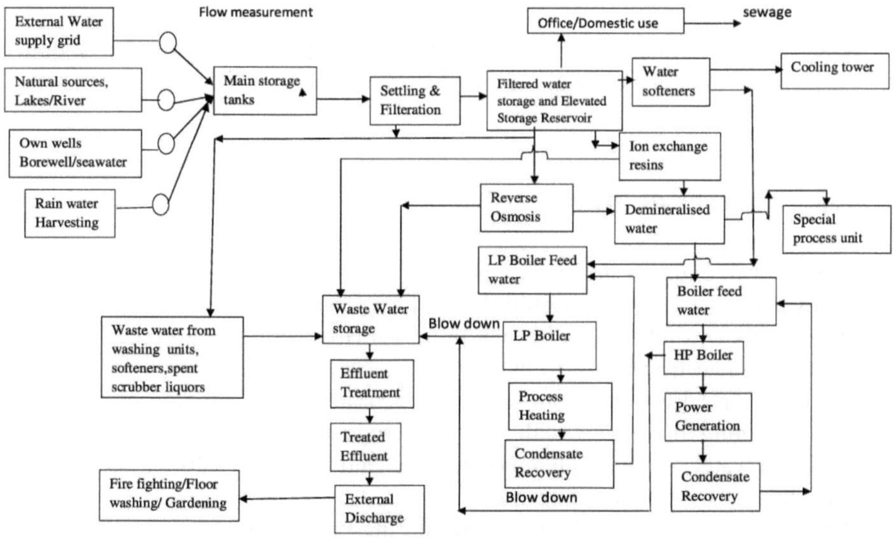

Typical water balance for a chemical process industry. *Notes* 1. Elevated Storage Reservoir is for supplying water in case of emergency to important units. 2. Process units can also be supplied Low Pressure steam (for heating) from tapping provided on Steam Turbine for Power generation

- For cooling of reactor jackets or as cooling spray on certain storage tanks;
- Water cooling of electrodes, glands for agitator shafts (of vessels handling inflammable fluids);
- As make-up of evaporation loss from cooling towers;
- As boiler feed water;
- Domestic use—drinking, toilets and washing floors, and gardening;
- Washing of reactors, tankers, process vessels;
- Cleaning of sludge in absorption towers and tower packing Raschig rings/Intalox saddles and internals, demister pads, candle demisters;
- Make-up of evaporation loss from scrubbing systems;
- Washing of effluent treatment units like lamella clarifiers, tube settlers, settling tanks, filter press;
- Safety showers, eyewash fountains;
- Testing of process pipes, storage tanks; and
- Fire fighting.

9.11.1 Quality of Water Required

The following parameters are generally analysed for the water samples drawn from raw as well as treated water streams to check suitability for the intended use.

Total dissolved solids, pH, suspended solids, temporary and permanent hardness (which can affect suitability *for process use*).

Conductivity, silica and dissolved oxygen content are important *for boiler feed water.*

Water treatment facilities required depend on the quality and quantity required for each of the intended uses in the plant.

9.11.2 Sources of Water

- Supply from external grid (state managed);
- Lakes and rivers (may be turbid during heavy rains);
- Open wells and tube wells in own premises (likely to contain more dissolved solids if drawn from deep wells);
- Sea water or salty water from sources which are not in common use;
- Water coming out along with gas and crude oil from oil wells (known as produced water)—It can contain oily sludge and dissolved salts and hence needs considerable treatment before use;
- Water harvesting—collection of water during rainfall and its proper storage;
- Recycle of treated effluent/drained out water from process units;
- Recycle of condensates from steam turbines, evaporators, process units being heated.

9.11.3 Storage Facilities for Water

All incoming waters are generally received in raw water tanks. These shall have partitions with drain valves and cleaning manholes. It is then possible to clean the sections one by one without disturbing the plant operations.

All tanks shall have level indicators, which are clearly visible from a distance, and low-level alarms. The overflow nozzles/lines from the tanks shall be connected to a header (which is then connected to another collection tank) so that no water is wasted at any time (water meters, flow indicators, and quantity totalisers shall also be provided to measure the quantities received into and consumed from important tanks).

All important water tanks shall be calibrated and have drain valves as well as cleaning manholes to enable thorough cleaning from inside (which should be done at least once a year). The lower exit nozzles shall be a little above the drain valves *so that settled matter will not go out with the treated water and will remain at bottom*. It can then be drained out when the tanks are cleaned from bottom manholes.

If possible, the lower portion of the tanks shall have conical shape for easy settling and removal of settled matter. This may, however, make a small volume of water *below the exit nozzle* unavailable for use (being a dead volume).

The water tanks in the process plant shall be designated as raw water and treated water tanks (filtered water, soft water, demineralised water, boiler feed water, dilution water, etc.).

Waste water collection tanks shall be located in such a way that all blowdowns from cooling towers, waste water from washing of reactors, process units, spent scrubbing liquor, etc., can flow by gravity into them. The waste water can then be either treated again (for reuse) or sent to effluent treatment plant for recovering as much water as possible before the final discharge outside the premises.

9.11.4 Elevated Storage Reservoirs (ESR)

These are generally provided in a plant to supply water by gravity to important points of use in case of power failures or when the water pumps are not working temporarily. They shall be conveniently located in a central place in the premises.

The ESR exit line should be connected to the main water lines in the plant through solenoid valves which open automatically (with arrangement to open manually if it does not open) when **power supply fails** or when pressure in main line falls below a set pressure. This is to maintain supply of water to all important points of use till diesel generator set is started for resuming power supply to important pumps and other units. It will give the operators the vital 15–20 min to take necessary steps till these pumps are restarted (on power supply from the DG sets—which will be gradually loaded).

Senior engineers shall decide *in consultation with experienced plant operating personnel* the order of importance of all critical water consuming units and the requirement *at each of them* for about half an hour. The ESR shall have a volume about 20 % more than the total of these individual requirements. Management shall invest an amount towards the cost of tank, its support structure, water piping for filling, automatically (opening) solenoid valves, etc.

Dedicated storages

Provide a totally assured supply of water for safety showers, eyewash fountains, fire fighting, and cooling water for spray on certain storages of inflammable fluids by having dedicated storage tanks of appropriate capacity at a height.

9.11.5 Water Treatment Plants

Raw water may contain floating impurities; organic impurities, dissolved calcium and magnesium bicarbonates, chlorides, and sulphates, as well as dissolved oxygen, etc. These can cause both temporary and permanent hardness. Untreated water can affect the product quality, cause corrosion of equipments, form scales in boilers (can consume excess fuel and affect their performance), and disturb process reactor temperature due to scales in the cooling jackets.

Following points shall be considered for selection:

1. <u>Quality of raw water</u> available from various sources
   pH, turbidity, floating impurities, total dissolved solids, presence of chlorides, nitrates, sulphates, bicarbonates. Select another source of water if the TDS is 450—500 ppm or more as it may be costly to treat this water. Raw water having less than 100–125 ppm TDS may be accepted.
   The source of water shall be selected on the basis of quality, quantity, reliability throughout the year and cost of obtaining the supply from the source to the plant (pipelines, canals, pumping from wells, by water tankers). Permission will be required to draw water from water storage bodies such as lakes, rivers, underground sources which are generally used for supply of water to general population by the state authorities.
2. <u>Quantity required for various purposes</u>:
   Treated water will be required as process water (for making solutions, as dilution water, as reaction medium) cooling water (for direct cooling of equipments and as cooled water from cooling towers), for boiler feed, flushing and washing of equipments, domestic needs, gardening, reserve stock for fire-fighting, etc. Hence, one shall correctly estimate the total expected demand (present and future) for the following and then arrive at the capacity of the water treatment plant required.

(a) Boiler feed water for high-pressure and low-pressure boilers.
(b) Quantity of condensate likely to be recycled for feeding again into the HP and LP boilers.
(c) Process water.
(d) Make-up water for cooling towers.
(e) Make-up water for refrigeration system (chilled water/chilled brine).
(f) Water requirement for plant washings and domestic purpose.

All such needs shall be correctly estimated for the present and the future.

3. The quality (analysis) of water required for these shall be known individually.
4. Operating cycles

   - Number of regeneration cycles/day: Usually one per day. (This can be two or more to reduce the cost of plant.)
   - Total quantity of treated water available per regeneration cycle in m^3.
   - Time required for regeneration and restart: 2 h (maximum). If more time is required, there is a danger of the treated water tank getting emptied.
   - Senior engineers shall decide whether the water treatment facilities will be operated round the clock (all 24 h of a day, 7 days a week) or for about 8–12 h only per day. This will need bigger units for filtration, softeners, ion exchangers, reverse osmosis plants, etc.

5. Essential components of the water treatment plant (those *depend on the process used for the treatment of raw water*):

   - Main storage for raw water;
   - Primary settling tank: If the water is muddy in rainy season;
   - Transfer pumps;
   - Pressure sand filter;
   - Active carbon filter;
   - Instrumentation;
   - Civil works; and
   - Electrical installation.

Water softener Ion exchange based: This is normally regenerated by sodium chloride, and thus, the treated water contains sodium salts.

9.11.5.1 Reverse Osmosis

This is suitable for treatment of water containing considerable amounts of dissolved solids. *The fraction of water rejected can be on the higher side (depends on the treated water quality desired) and one should check if there is any problem at site for disposal of the wastewater generated due to backwashing of sand filter, active carbon filter, resin beds* (can it be reused in the premises, for scrubbing, gardening, fire fighting?).

Permission shall be obtained for this from statutory authorities before procuring this type of plant. Also one shall look into the reliable availability of the membranes being used, their life, and the power consumption per cubic metre of treated water.

Following items shall be included in the supply of RO plant by the vendor:

- Sand filter (one working + one standby);
- Active carbon filter (one working + one standby);
- Iron removal filter (one working + one standby);
- Micron cartridge filter (one working + one standby);
- Degasser;
- Iron removal unit;
- Anion and cation resin columns;
- Mixed bed polishing unit;
- Acid and alkali tanks for regeneration;
- Air blower with motor;
- Acid and alkali pumps;
- Membrane blocks (allowable maximum pressure to be specified).
- High-pressure pumps shall not generate pressure which may damage the membrane;
- Deaerator for removing dissolved gases;
- Oxygen scavenging (dosing tank and pump for chemical addition like hydrazine) *Consult boiler manufacturer if necessary*;
- Pressure gauges;
- Online conductivity meter with remote digital indication;
- Rotameters and pH meters for treated water;
- High-pressure piping shall be of stainless steel 316 L; and
- Low-pressure piping shall be of a material which will not contaminate the treated water (CPVC with test certificates).

9.11.6 Specifications of Treated Water

- Process water: Filtered, free from organic and suspended matter, pH = 7.0–7.5, TDS less than 100 ppm, hardness less than 50 ppm.
- Soft water (for LP boilers): Operating below 20.0 kg/cm^2, hardness below 10 ppm, pH = 7.5–8.0, TDS below 10 ppm as far as possible to reduce blowdowns.
- Demineralised water (for HP boilers): Generally operating above 15.0 kg/cm^2, TDS $<$ 1.0 ppm, pH = 8.5–9.0, hardness $<$ 0.2 ppm, silica $<$ 0.1–0.2 ppm, *conductivity (mho): as specified by the boiler manufacturer.*

Chapter 10
Instrumentation

10.1 Introduction

Proper instrumentation is necessary for safe and efficient working of a chemical process plant since it is necessary for the operating personnel to know how to correct the flow, temperatures, pressures, etc. of the process streams. It calls for accurate measurement, recording and controlling of key process parameters.

10.2 Main Objectives

10.2.1 For Operating Units

- Safety of personnel inside and outside the premises who may get affected due to accidents in the process plant. Safety of process units and machinery shall also be ensured
- Minimise generation of effluents
- Proper treatment of the effluents for environmental pollution control
- To maximise energy efficiency (minimise consumption and maximise recovery)
- Maximise production from process plant while minimising consumption of raw materials
- To maintain product quality as per specification (to minimise the rejection of products)
- Minimising machinery breakdowns by running the equipment under control
- To maintain delivery schedules of required products as per market demands.

Research and Development activities (for new process and technology) also need good-quality instruments that should give accurate results.

K.R. Golwalkar, *Production Management of Chemical Industries*,
DOI 10.1007/978-3-319-28253-4_10

10.2.2 For Safety of Personnel and Plant Equipments

- Timely warning of unsafe conditions such as high pressure or temperature in process plant system; formation of explosive condition due to extinguished flame in a furnace.
- Overheating of material inside the process reactors
- Low level in overhead tanks (for safety showers, for fire fighting)
- Low level in feed water tank for boilers
- Low pressure in cooling water supply lines
- Toxic gas concentration in working areas or closed vessels (which must be checked before going inside for cleaning or any maintenance work).

10.2.3 For Pollution Control

- Ensuring proper operation of process plant to minimise effluent generation
- Tracking of pollutants, such as SO_2, acid mist, and particulate matter in stack gases, low pH, concentration of dissolved solids, and suspended particles in liquid effluents.
- Warning operators by audio–visual alarms if there is possibility of environmental pollution due to the presence of excess pollutants in exit gases (warning for abnormal gas analysis of flue gases)
- Automatic pH control, aeration for treatment of effluents, and automatic recycle for additional treatment (if necessary) for properly neutralising pollutants.

10.2.4 For Energy Efficiency

- Monitoring power consumption by individual drives, CO_2 % in exit gases (fuel consumption of burners fitted to furnaces), power required for air blowers (indicates system choking), and high surface temperatures of equipment shells and ducts (indicates damage to inner thermal protective lining);
- Optimising operation of waste heat recovery units, refrigeration systems, preheaters for air and water, cooling tower fan, compressed air systems, etc.
- Monitoring performance of steam consuming units such as process reactors, multiple effect evaporators: (*measuring their steam consumption and condensate recovery*);
- Monitoring the draw of power from external grid; own generation by turbine-generator as well as DG sets.

10.2.5 For Ensuring Quality of Products

- Checking the products for purity, moisture content, insoluble matter, and free acidity to ensure customer satisfaction.
- Monitoring the process parameters while operating the plant as per standard procedures laid down.
- Ensuring smooth process plant operations by quality/analysis checks at every stage of manufacture, e.g. analysis of raw materials at the time of procurement, progress of reaction, activity of catalysts, refinement of crude products into acceptable grade by controls on flow, temperature, and concentration.
- Reducing equipment breakdowns by better process control and timely alarms for abnormal situations.

10.2.6 For Efficiency of Operations

- Checking quality of raw materials *on arrival and after treatment for purification* before feeding into reactors and other process units.
- Minimise consumption of raw materials by (i) measuring and controlling feed rates; (ii) measuring process conditions at every stage; and (iii) analysis of exit streams.
- Minimise consumption of utilities by optimising operation of process units and water treatment plants, the use of compressed air by automatic control of air compressor operations, and the operation of refrigeration systems.
- Automatic control and warning of abnormal feed rates of reactants and high temperatures in process units, excessive flow rates which can disturb catalyst beds and conversion efficiency, affected production rates, or waste reactants due to incomplete conversion.
- Early detection and preventing overloading of reactors, catalyst, and plant equipments.

10.3 Types of Instruments for Process Monitoring

The process shall be monitored by careful selection of the following typical instruments generally required in a chemical plant.

Additional instruments shall be installed as per need.

- Level indicators, transmitters, and controllers
- Thermocouples, resistance temperature detectors, and controllers
- Pressure indicators, transmitters, and controllers
- Flow rate indicators, totalisers, and controllers

- Toxic gas detectors and stack gas analysers
- pH sensors and controllers
- Electrical measurements and monitoring of boiler operation.

All these instruments shall have in-built recorders or shall be connected to a centralised computer for recording the data. The information shall be retrievable whenever required to know the past records and the trends over a suitable time period.

10.4 Selection Criteria for Plant Control Instruments

- Maximum and minimum values of parameter to be controlled;
- Suitable range shall be chosen for the measurement—the usual value shall not be towards the end of instrument range;
- Set points shall be easily adjustable;
- Dials and figures shall be easy to read;
- Should have the desired accuracy for the application;
- Location in process equipment whether inside the unit or outside the surface;
- Material of construction of sensor, and wetted parts of the instrument shall not be affected by the process fluids;
- The probe and actuators for controller shall not get affected by process fluids, dust, gases, and vapours present in the vicinity. Protective cover shall be provided by vendor. *The probes shall not be susceptible to vibrations;*
- Actuators can be electronic, pneumatic, hydraulic, or mechanical. This shall be discussed between vendors and plant engineers. Flame proof fittings shall be provided if inflammable/toxic, explosive vapours are nearby;
- Type of display (Digital or Analogue) shall be as required by operating persons;
- Instrument dial shall be linear and easy to read (the indicated value);
- Dial shall be backlit-type, and sufficiently big size for ease of reading;
- Auxiliary contacts for data transmission/alarm signals/recording of trends and data retrieval shall be available;
- Response shall be quick;
- Plant engineers shall not use very long sampling lines for fluids. Truly representative sample shall be drawn from process/from inlet and exit of effluent treatment plant for correct results in real time;
- Long sampling line can delay measurement of parameter and subsequent control action (time for getting result of sample = volume of line/rate of draw of sample) + time required for analysis;
- The set points shall be adjustable and shall not drift by itself;
- Shall be multifunctional—shall indicate, record, control, and warn in dangerous situations;
- Instruments shall be calibrated by the manufacturer before supply;
- Convenient to repair. Spares shall be immediately available;

- Affordable initial cost;
- Short delivery period; and
- Explore if an annual maintenance contract is possible with vendor.

10.5 Installation and Commissioning

This shall be preferably done by the vendor and shall be included in the purchase order. Vendor shall also supply spares for two years along with the instrument.

- Obtain copies of all instruction manuals, list, and specifications of spares from vendor;
- Plant management shall fulfil all obligations and comply with site conditions required as mentioned in the guarantee cards. (e.g. sample coolers/conditioners and strainers in sampling lines to protect delicate sensors);
- Online instruments should have isolating valves and a bypass line to continue operations when the instrument is under repair;
- Vendor shall provide proper weatherproof and dust-proof cover. Plant engineers have to arrange for steady power supply at required voltage and frequency as per manufacturer's recommendation;
- Fabricate comfortable work platform for the instrument, connecting pipes, and cables;
- Arrange shadowless lighting for observing the dials of field instruments and for maintenance on the spot if required;
- Connecting cables shall not be too long, shall not pass through hot areas, strong electrical or magnetic fields, or directly below pipelines carrying corrosive chemicals.

10.6 Some Practical Suggestions

It is seen that in spite of taking precautions for the proper installation and careful operations, some of the instruments and controls do not read/control correctly. In such cases, the inferences drawn for the process conditions or equipment could be erroneous.

Some typical reasons for this are given below along with some small practical suggestions for the consideration of the plant engineers. The aim was to assist the plant engineers for plant operations in case of a malfunctioning of the instruments or controls. They are requested to look into these and carry out the modifications/implement in their plants *if found suitable*. The inferences can be drawn by observing other equipment and process conditions in other units in the plant also, instead of relying on one particular instrument only.

One shall look objectively and logically at the prevailing situation in the plant.

10.6.1 Float-Type Level Indicator

A float moves up and down in a guide tube as per level of liquid in the tank. A wire is attached to the float, and an external pointer (corresponding to the position of the float) indicates the level on the outside scale. In case of dangerous liquids, the float has a straight rod fixed at the top with a strong magnet. This rod moves up and down in a stainless steel tube which has an external follower magnet and a graduated scale. The level of the liquid inside the tank is indicated by the external follower magnet.

Some typical problems which may come up:

- Float gets stuck in sludge at bottom or in the guide tube;
- Connecting wire with the float gets broken;
- Mismatch between float position and external pointer moving on scale;
- Connecting wire comes off pulleys;
- Drain openings in guide tube get choked; and
- Float gets punctured and sinks.

10.6.2 Suggestions

- Guide tube shall be removable from tank. Clean the tube regularly.
- Consider small diversion of incoming flow to the tube also for self-cleaning of the tube and drain holes at bottom.
- Provide reference nozzle at top. Measure empty space from the reference nozzle by a dipstick in case of urgency. Take care to wash the dipstick if acid level is checked.
- Material of Construction of Float shall be corrosion-resistant Stainless Steel 316 or carbon steel coated with Teflon or glazed ceramic.
- Use strong multistrand wire (made of corrosion-resistant material) for connecting to the float.
- Provide lubrication and guide clips at pulleys so the wire will not come off.
- Replace float after every two years at least.
- Provision of external transparent tubes (toughened glass or polyethylene) on storages tanks fixed on small taps with isolating valves. These are special valves with a small ball inside. It can fit tightly on the seat so that in case of damage to the tube, the ball closes the opening in the seat and no liquid can come out.
- Tankers for liquids: Sight glass can be provided in addition to the level indicator.

10.6.3 Non-functioning of High-Level Alarms

If there is malfunction of level indicator or high-level alarm and filling is continued in the tank, there is a chance of overflow (spillage) outside. It is advisable to connect overflow nozzles of all tanks to prevent spillage from the tank which is being filled. The liquid from the tank flows into another tank. This can prevent accidents. *However, if the liquid overflows into a tank from which material is being fed to process plant, it will become difficult to determine the consumption correctly* (see Fig. 6.1 in Chap. 6).

Another arrangement is to connect the overflow nozzles of all the tanks to a common header at a little lower level. This shall be connected to a **spare tank** which shall be always kept empty. Plant engineers may choose between these two arrangements as per their convenience.

10.7 Ultrasonic Instruments for Level Indication

- Dusty atmosphere in vessel may distort the return signal to the sensor;
- High temperature of materials can change the speed of signals; and
- Ladders or internal material build-up may cause indirect reflection from sloping surface leading to weak and/or split echoes of signals.

Fix convex mirrors above a deep tank to observe the amount of solid inside from the ground level if it is not possible to go up and take measurement by a dipstick, if the ultrasound detector is not working properly, or the deep tank is at a height.

10.8 Pressure Gauges

- Pressure taps gets choked due to the deposit of ash or dust.
  Clean the taps regularly or inject inert gases such as nitrogen into the tap from a tee connection once or twice in a shift to keep it clean.
- Pressure gauge shows fluctuating reading—can be due to flap in butterfly valve fluttering inside. Check the position of flap and tighten the lock nut.
  Provide seal pots/vents at appropriate places to release excess pressure if the released material is not dangerous or harmful to environment.

It is better to be on safer side while operating units under pressure. Hence, care shall be taken:

(i) not to feed excessive amounts of reactants(feed only measured amounts from small day tanks by means of gear pumps/VFD run motors) or
(ii) not to increase temperatures too fast (strict control on heating medium)

(iii) design and fabricate the vessels strictly as per ASME Sec VIII standard—*more details are given in appendix on pressure vessels*
(iv) provide safety devices as per statutory directives.

10.9 Thermocouple/Dial Thermometer

The readings could be erroneous due to the following reasons:

- Burner flame is impinging directly on temperature sensor. Thermowell is damaged due to flame. Check burner installation, tertiary air distributor, and atomisation of fuel oil which can cause this problem. Provide additional view glasses on furnace shell to observe the flame. Clean strainer in fuel oil pipe.
- Low thermal conductivity of ceramic thermowell slows the response. Use heat-resistant stainless steel for fabricating the thermowell.
- Very low level of oil in the thermowell for dial thermometer—(hence check the oil level regularly).
- Improper location of thermowell/sensor.
  It should be in the flowing gas/liquid and shall not be in an idle pocket of the system.
  In case of converter of a sulphuric acid plant, it should be at interface of catalyst bed and quartz bed and not inserted in the catalyst bed itself.
- Sensor is damaged either at tip or in between—replace by another one at the earliest.
- The compensating cable is cut/the contacts are corroded—check continuity
- Rainwater falling on contact of connecting cables with thermocouple can cause errors (hence provide weatherproof protective cover for junction point).

10.9.1 Watching Carefully

The following shall be watched carefully if the temperature readings are doubtful.

- Check feed rates of fuel pumps, current drawn by the drive motors, setting on the variable frequency drive if provided. *Use gear pumps for fuel oil as far as possible.*
- Monitor the drop in level of fuel oil feed tank every half an hour to determine the actual quantity of fuel consumed.
- Exit flue gas composition can indicate excess or low feed of fuel oil (or combustible reactants such as sulphur, ammonia). Provide online analyser for CO2 %, free oxygen %, etc. or check gas analysis by Orsat apparatus. Test exit gases for sooty particles by clean filter paper for any excess oil feed.

- Observe the temperature of downstream units or pipelines which may rise abnormally; there could be discoloration of the paint on external surface or it may even get peeled off from the surface. This indicates excess fuel oil supply.
- Observe the colour and length of the flame inside—orange smoky flame indicates excess fuel supply; bright yellow incandescent flame indicates high temperature and proper combustion.
- Observe the production rate of the plant (when ammonia or sulphur is being burnt), and the rise or fall in temperatures of downstream units which depend on the feed rate of these reactants.

10.10 Rotameters

- Suspended particles in fluid can disturb movement of float;
- Higher temperature of fluid can evolve dissolved gases in fluid (and disturb movement of float);
- Change in density or viscosity of fluid at higher temperature;
- Deposits on transparent tube make position of float difficult to observe;
- Float has become lighter as it has worn out; and
- Float has become heavier due to deposits.

10.11 Turbine Metres

- The blades have deposits or are worn out;
- Accuracy can be affected by the evolution of bubbles in fluid;
- Rotor *may not rotate properly* if fluid viscosity is high or flow is laminar;
- Rotor movement can get obstructed by suspended solids also. Needs a good strainer upstream.

Used only for measuring clean, steady, medium- to high-speed flow of low-viscosity fluids.

10.12 Vortex Metre

- Working principle—An obstacle is placed perpendicular to the axis of pipe in which the fluid is flowing. Eddies are generated across the sides of the obstacle which are detected by electronic sensor. Their frequency is proportional to the fluid flow;

- Not very suitable for low flow rates (sufficiently measurable eddies may not be created);
- Worn out/shaky impingement obstacle;
- Problem in electronic sensor/circuit; and
- Very low flow rate of fluid.

10.13 Some More Suggestions

Select only appropriate flowmeter as per range of flow rate and fluid properties. Provide strainers before the flowmeters and clean them regularly. The screens shall be of corrosion-resistant material and may be replaced every year or earlier if necessary.

The connecting cables for electronic parts must be well protected from dust, rain, or any corrosive liquids which may fall on them from a leak (if the pipeline carrying such liquid is laid directly above the connecting cable).

If the indication by the flowmeter is erroneous, it can result in excessive feeding of raw material resulting in the development of high pressure or temperature in process units, or overflow of downstream process tank.

Use metering pumps instead of centrifugal pumps since the latter can feed in excessive amounts of reactants, fuel oils, and acids. Metering pump is more suitable for controlled feeding.

The discharge flow of these pumps is as per setting made (*which shall be locked and recorded in the operation log book every shift*) and is always constant even with an increase in temperature or viscosity, presence of suspended solids, etc.

Feed from small, calibrated day tanks for feeding liquids. The capacity of these tanks shall be enough for running the plant only for about 2 to 4 h. The level in these tanks shall be monitored every hour. This precaution can minimise the chance of overfeeding of liquids. This can prevent process and safety problems.

While feeding liquids by gravity flow, it shall be through rotameters and provide bypass lines only if necessary. If metering pumps are to be used, the discharge line of the pump shall first rise above the feed tank and then connect to the process unit. This will prevent flow of liquid by gravity (when the pump trips—but valves in suction and discharge lines may be open when the pump has tripped).

10.14 For Feeding Liquids to Small Process Units

Feed liquids into small units by containers of 10/20 l capacity and counting their number. This will prevent excess feeding *if a pump is used* for feeding from a tank.

10.15 Gas Flowmeter Reading Incorrect

10.15.1 Precaution—Use a Roots-Type Blower with Variable Frequency-Driven Motor

A centrifugal blower shall not be able to deliver steady constant volume of gas if there is a choke in the system and a high backpressure has occurred. This can seriously affect plant performance if a constant volume is required to maintain gas stream concentration or to prevent explosion in a furnace due to excessive fuel vapour concentration. It can be misleading if the flowmeter in airline is also not indicating correctly and the situation can get aggravated. *The motor connected to the centrifugal blower may not draw heavy current in such a case.*

However, the motor for roots blower will draw heavy current if there is a high backpressure due to some choke in the system and will continue to deliver air. It will not matter whether the air/gas flowmeter is working accurately or not.

Provide a revolution counter metre, i.e. RPM metre, for the rotational speed. The roots blower delivers an almost steady volume (the desired value can be set as per speed)

Provide manometer/pressure gauge for gas pressure measurement.

The valve on discharge side of the blower shall have a locking arrangement to keep the position fixed. Butterfly valves in gas ducts shall have an observation window for checking the actual position of flap. This also needs to be monitored.

Example: Manufacture of Sulphuric Acid

If the airflow reading is doubtful, then check the temperature and SO_2 % of exit gases from the furnace, operation of downstream units, and production rate of the plant. Check setting on metering pumps, current drawn by blower motor, temperatures of furnace and downstream converter passes, and rate of dilution water added, and correlate to material balance once a day/two days to estimate the loss of unreacted SO_2 from stack.

High exit gas temperature from furnace indicates **low** airflow rate. This is confirmed by **higher SO_2** %. However, if the production rate has not increased corresponding to the high **airflow rate indicated by the air flowmeter, it** is further confirmation of **low** airflow rate. Less consumption of dilution water *may not be relied upon* because it depends on the prevailing relative humidity.

10.16 Some Practical Tips if Thermocouples are Showing Incorrect Temperature Readings

- Monitor the surface temperature of refractory lined furnaces to get an indication that internal refractory may be getting damaged or excessive fuel is being fed into it. This shall be done regularly at the same spots on the shell chosen at

one-metre intervals along the length. Three spots shall be marked at angles of 0, 90, and 180° on the circumference at the same cross section.
- Pieces of paraffin wax candles and solder metal can be used for approximately detecting whether the temperature of the surface is below 100 °C.
- Sprinkle a little water on the external surface of the process unit or the gas duct. If it boils off immediately, the temperature is more than 100 °C. Now draw a line on the surface with a piece of solid sulphur—if it melts immediately, the temperature is more than 120°; if it ignites, it is 250 °C. If a line is drawn with a piece of lead melts, then it is 330 °C. Such surface temperatures definitely indicate damage of internal refractory lining (melting point of tin is 232 °C and zinc is 419.4 °C. These can also be used).
- Any discolouration of ducts, or the paint on surface of duct getting peeled off can be due to high temperature of furnace and exit gases.
- Provide high-temperature alarms for furnace exit gases if required (when burning fuel oil).

10.17 Gas Leak Detection

10.17.1 A Word of Caution

These practical methods are crude. If they are not able to detect a leak, it will not mean there is no leak and it is safe to move around in the plant.

- Test papers for NH_3, H_2S, SO_2...Cl_2 can indicate some leak of these gases in a plant though they will not be able to measure the concentration.
- A wet piece of carbon steel piece kept hanging near a duct can become blackish if there is H_2S gas leak.
- A wet swab with HCl will start fuming in case of an ammonia NH_3 leak.
- Similarly a burning piece of sulphur will also start fuming in case of an ammonia leak.

10.18 Incorrect Readings by Weigh Bridge

- Use calibrated metering tanks for sale of liquids (acids, organic fluids). These shall have a sight glass and overflow line back into main storage tank at known capacity. A pump is run to fill up the metering tank. When it gets filled up to the desired capacity, the excess liquid flows back into the main storage. Now the metering tank can be emptied into the dispatch tanker. Two or three overflow nozzles can be provided on the metering tank as per different desired capacities. This arrangement can deliver correct quantities of products even when the weigh bridge is not working properly.

10.19 Strength Indicator for Sulphuric Acid (Indicating Low Strength But Actual Strength May Be High)

- Watch exit gas appearance—curly dense fumes of SO_3 mean high strength of acid. A match stick dipped in a sample of acid will catch fire instantly if the strength is more than 98 %.
- Dilute a known amount of acid sample with water and observe the rise in temperature. Refer to standard charts to know the strength of acid.

10.20 Stack Gas Analyser for Sulphuric Acid Plant

Comparison of thermal conductivities of gas stream is done with gas mixture of known strength. It may show erroneous results due to the breakdown of suction blower, choking of sampling probe or conditioner, leaking connecting tube, and fault with electronic circuit of indicator and recorder or exhausting of the chemical standard solutions used for titration.

It is suggested to:

(i) Confirm by manually checking gas samples once/twice a shift.
(ii) Replace standard solution in the instrument regularly.

10.21 Stack Gas Analyser for Absorbed Exit for HCI Gas

- A quick test by a wet swab dipped in liquor ammonia solution can be done by holding it against the exit gas to detect the presence of unabsorbed HCl gas.
- **Quick check of scrubbing liquor**—Provide a sampling line from pump discharge to control room and check by pH papers or universal indicator solutions. This is to check the alkalinity of scrubbing liquor if the pH metre is not working.
- **Solid feed measurement**
  Belt conveyers, screw conveyer, and bucket elevators are used for feeding solids such as rock phosphate, coal, salt, and bauxite in large quantities. This feed system can be calibrated by operating them by feeding known amount from a stack/heap of these materials once a week. *Stacks of known amounts* (*20 tonnes, 50 tonnes,* etc.) *shall be arranged at the feed point and time required for conveying the entire amount can be noted. Care shall be taken to see that the trough of the screw or the buckets of the bucket elevator are always full.*

10.22 General

10.22.1 Checking Purity of Filtered Liquids

This is tested in a laboratory generally and it takes considerable time to do so. A quick test is to dip a piece of thick bright white paper sheet into the filtered liquid. Presence of black spots on the paper sheet will indicate unsatisfactory filtration and need for recycling it. This can be used for checking quality of filtered sulphur coming out from a pressure leaf filter.

10.23 Feed Control by Centrifugal Pump

If a centrifugal pump with high discharge capacity is being used to feed a process unit with manual operation control, then provide a recirculation line (to the feed tank on inlet side of the pump) with an orifice plate and a throttle valve.

The quantity to be fed to the process unit can be controlled by the recirculation of the liquid into the feeding tank. The throttle valve can be operated by automatic control with a manual override.

10.24 Calibration Test Benches

Make provision in plant premises to immediately attend and recalibrate faulty instruments.

Chapter 11
Energy Efficiency

All process equipment and machineries as well as installed facilities shall be designed for maximising the energy efficiency (by minimising energy consumptions and maximising energy recoveries). This is necessary to reduce the cost of production. The chemical plants require mostly thermal and electrical energies. Other types of energy—mechanical (gravity heads, pressure heads)—are also used to a small extent. *Optical or ultraviolet energy is used for special reactions, but is not addressed in this book.*

Selection of the proper process and technology is very important for achieving the aim of energy efficiency.

11.1 Thermal Energy

Energy balance of the plant is the statement of all energy inputs to the plant and outputs from the plant. The inputs shall include the need to heat the incoming materials for preliminary processing (e.g. *drying, melting, calcination*), for carrying out endothermic reactions, and for collecting and processing the product. A study of these matters will enable estimation of required energy for the plant.

Energy output—A careful study of the thermal energy evolved during exothermic reactions as well as all the energy going out with hot streams shall also be carried out. This shall be balanced against the total energy input to know the net energy required for running the plant. Any additional requirement of energy above the calculated amount may be due to the losses of energy from the system. The reasons for losses shall be found out and attempts undertaken to minimise them.

It may become possible to reduce the losses by modifications of the process, equipment or operating methods such as saving energy by *modification of process technology, handling lower volumes of reactants, better refractory lining*, or recovering energy by *preheating of incoming dilute solution to an evaporator by hot outgoing concentrated solution from the same evaporator*. Engineers shall always be on the lookout for such possibilities.

An example from a sulphuric acid plant is as follows (*modification of process technology*):

K.R. Golwalkar, *Production Management of Chemical Industries*,
DOI 10.1007/978-3-319-28253-4_11

When the air blower is installed **before** the drying tower (DT) in a sulphuric acid plant, the heat of compression of the air gets added to the circulating acid in the DT. But it is lost when the acid is cooled. However, *installing the air blower **after the DT** in the plant can save this loss. It becomes possible to use the heat of compression of the airstream as the air gets heated up and enters the sulphur burning furnace at a higher temperature. It is then recovered as additional steam generated by the Waste Heat Recovery Boiler.*

11.1.1 Modifications for Reducing Energy Consumption

Process modifications:
In case of sulphuric acid plants, the earlier single-conversion single-absorption process was replaced by the double-conversion double-absorption process for most of the new plants. The SO_2 % in gases at converter inlet was increased from 7.5–8.0 % to 11.0–11.5 % which could reduce the volume of gases to be handled for the same production rate. The power required by the air blower got reduced as a result.

Equipment modifications:
Provision of an additional layer of insulating bricks inside a furnace can reduce the temperature *of the outer surface of the shell.* This can reduce the loss of heat from the outer surface of the furnace shell.

Provision of external cladding by aluminium sheets on the existing mineral or glass wool insulation *in a leak-proof manner* can further reduce the loss of heat from surfaces of equipments (reactors, heat exchangers) and hot ducts (larger-diameter pipes).

Replacement of water-cooled steel electrodes by graphite electrodes in electric furnaces used for carbon disulphide manufacture could reduce power consumption by about 8–10 %.

Modifications in Procurement of materials (*in a ready/convenient-to-use form*):
Obtaining alum as a ready-to-use solution at the point of use (instead of solid lumps) saves energy for dissolvers. It also saves energy required for evaporation of the dilute solution at the supplier's end. This is successful when the distance for transportation of dilute solution is very less.

Procurement of molten sulphur (instead of solid lumps) for a sulphuric acid plant will reduce the steam required to melt the solid sulphur.

Obtaining small carbouys (which are filled with known weight of material like stabilisers) instead of filling from main storage tanks. These small carbouys can be charged into process units directly. This can save time, effort, and energy for such charging operations.

Natural gas is available at a pressure from oil wells, and its supply can be obtained at point of consumption by pipelines (with appropriate safety precautions and metering system). This method of obtaining fuel can save efforts and the energy required *for transportation, unloading at site and then pumping to points of use* in the case when fuel is obtained as furnace oil filled in tanker loads.

Careful Operations, Maintenance, and Modifications:

Manufacture of Alum: It needs steam for heating the reaction mass. The consumption of steam can be reduced by adding water and concentrated sulphuric acid *initially* to the reactor. The heat of dilution can raise the temperature of the reaction medium. Steam is to be supplied *only later* in small amounts for completing the reaction with alumina hydrate.

Boiler operation: The exit flue gas/hot process gas temperature is always considerably above the saturation temperature corresponding to the pressure of the steam at which it is being generated. A study shall be carried out in the plant to ascertain the pressure at which steam is really required. The boiler shall be operated at a little above this pressure only. This reduces loss of heat if the economiser or the air preheater is not efficient to recover heat from the gases (could be due to ash deposit or corrosion of the tubes).

Evaporator operation: Deposition of salt deposits on heat transfer surfaces (tubes, jackets, coils) shall be minimised by controlling concentration of solutions.

Absorption towers (Gas–liquid contact operation): Consult manufacture of tower internals and packings which have maximum contact area for gas and liquid. Also operate the system at optimum incoming and outgoing absorbent/gas concentration and temperatures. Increasing the packing height in the tower may also help. This can reduce the requirement of circulation rate of the liquid and will thus save power consumption.

Bucket elevators, belt conveyor, and pneumatic conveying system: These shall be run only when the level in silos at delivery end goes down to a preset level. Instruments shall be provided to monitor the levels; provision of convex mirrors shall also be made to observe the levels (of open silos). The conveying systems shall never run idle, or insufficiently loaded. Ensure that flow of material is smooth at feed point (e.g. all buckets are properly filled up; there is no spillage of materials from belts).

Thermal insulation: All hot insulation shall have optimum thickness (excessive thickness can cause more heat loss due to extra surface area exposed), and the cladding by aluminium sheets must be leak-proof—no ingress of moisture or rainwater shall be possible. Likewise, no hot gases or heat from any source shall be allowed to enter a refrigerated system.

Use of chilled water or brine: It shall be minimised by providing external cooling fins on ducts, using cooling tower water. A temperature monitoring system shall switch off the cooling tower fans when the water is sufficiently cooled (especially during winter).

Proper lubrication and maintenance of gearboxes, tensioning of belt drives, and repair of agitator blades are some more measures which can reduce energy consumption.

11.1.2 Better Heat Exchanger Designs

Disc- and doughnut-type heat exchangers for gases have lesser pressure drops and do not have idle gas pockets (which may occur in conventional shell- and tube-type heat exchangers with segmental baffles), and the orientations of inlet and exit nozzles for the process streams can be made *to suit the ducts* (routes of gas duct can also be shortened which is helpful to reduce the pressure drops) without generally affecting plant performance.

These features increase energy efficiency of the plant since air blower needs less power, and gas-to-gas heat exchange is improved.

Plate heat exchangers can recover heat from hot sulphuric acid as hot water (*earlier designs of trombone-type acid coolers wasted all the heat*). The hot water can be obtained at a temperature within three to five degrees of the temperature of the incoming hot acid stream and hence can be used elsewhere in the premises.

11.2 Changes in Plant Layout

Reduce movement of materials to and fro, or up and down while designing feed systems by hoists or EOT cranes to reactors and process vessels.

The crude product shall be taken out from a nozzle facing purification units, and the output of purified product shall be taken out from nozzles towards storage tanks/loading points for shipments.

The plant shall have a centralised maintenance shed if possible, and it shall be as near to all process units and machineries as possible.

11.3 General Energy Saving Methods in the Plant Systems

Water supply

Store water in overhead tanks (with automatic level controller to start and stop the water pump) for use in many sections of the plant. This will reduce continuous running of the main water pump.

Use larger-diameter pipes for transferring liquids over longer distance to reduce pressure drops and hence the power required for pumping the liquid. The combined total of the cost of pipelines (along with supports plus other fittings); and the expected cost of power consumption is to be calculated for different pipe diameters. Note down the diameter of pipe corresponding to the *minimum total* of the capital plus operating costs. Choose this diameter or the next *higher available diameter* in the market.

This can also help in optimisation of pump capacity for the present and future needs of the plant (if next higher diameter is chosen) and efficiency of pumping (lower pressure drops).

The cost of valves, bends, and flanges may appear more initially—but will be found to be of advantage in the long run.

Compressed air

The loading and unloading cycles of the air compressor shall be set according to the capacity of air receiver and the pressure and volume of compressed air required at various points of use. Avoid long distances between air receiver and point of use.

This can be done by a study of the air lines, and after consulting the manufacturer, a suitable model of the compressor can be bought to minimise the power consumption.

The air receiver shall be designed and fabricated for a pressure sufficiently higher than the system pressure (normally used pressure in the plant system).

Safety valves on the air receiver shall be set as instructed by the statutory inspection authority so as to ensure safety of all personnel. Minimise frequent unnecessary blow off from safety valves (which can waste compressed air and hence power also). Regularly monitor condition of compressed air pipes and immediately attend leaks.

11.4 Reduce Energy Consumption for Material Transfer

This can be done by less handling of materials when the plant layout is designed accordingly.

- Incoming material shall be unloaded and stored as near the pretreatment facilities as possible. These, in turn, shall be near the feeding points of the process units. Further movement of the materials shall be by gravity as much as possible instead of using pumps, conveyors, or elevators.
- Direct dumping in point of use
  Reaction mass from first reactor is sent to second reactor by gravity flow and likewise from second reactor to third reactor during manufacture of alum by a continuous process. It then flows into the settler/clarifier and finally to evaporator by gravity.
- The packing and weighing sections for the finished products shall be nearer to the loading point of trucks (railway cars) for final dispatch to customers of the product.
- The waste treatment facilities shall be near the points where most of the ash, waste materials, and effluents are generated.
- Buying ready-to-use pulverised materials (feeding to reactors, charging into dissolvers) in bags or containers having standard weight can reduce energy spent on crushing and grinding.

- Obtaining fuel gas from external grid operating under pressure (*this will not need a separate gas compressor for feeding to furnace*).
- Obtaining raw water from external supply lines operating under pressure and directly feeding into filters/treatment units (*this will not need a separate pump for feeding to filters*).
- When the ducts and pipelines have larger diameter, the fluid velocity reduces, and consequently, the pressure drop for the same flow rate also reduces. This is also the case when the process units have more cross-sectional area available for flow. Power consumption can be thus reduced by a choice of larger cross section. However, the cost of the pipelines and the equipments may increase with higher size. An optimisation shall be done by considering the total of initial capital cost plus the power cost.
  Example: Use of gradually increasing diameter of gas ducts at entry of heat exchanger can **reduce entry losses** as compared to the loss when a small-diameter duct is directly connected to the heat exchanger (due to the formation of vortices).
- Effect of larger cross section/reduced fluid velocities shall also be considered on other process parameters, conversion of reactants, and reduction in coefficient of heat transfer due to reduced turbulence before taking a final decision.
- Better power transmission by V belts instead of flat belts—the V belts fit better in the grooves of the pulley. There is less slip, and more power can be transmitted.
- Hydraulic motors instead of high-torque electrical motors—this system uses high-pressure hydraulic oil to run agitators in reactors with very viscous materials which may need electrical motors of much higher rating.
- Power recovery from high-pressure exit fluids/gases—(exit gases from nitric acid plants are at pressure and are used to run power recovery equipments).
- Use of flywheels for using kinetic energy in crushers/grinders. The sudden jerks on the machines are reduced which can cause sudden heavy load on the drive motors.

11.5 Saving Electrical Energy

Electrical energy is a very important and versatile source of energy. It is easy to use for many different applications such as heating, material transfer, and electrolysis. Process parameters can be easily controlled (temperatures, flow rates, reactions, etc.). However, it is costly and needs properly designed set-up for receiving and using it safely in the plant. Efforts shall be done to see that the installations do not break down and to minimise power consumption. The following tips will be found useful by the plant engineers:

Arrange proper distribution of electrical load in different phases so that no phase is overloaded. This is necessary to reduce imbalance on transformer windings (and overheating in some).

Provide motors of optimum horsepower and torque if viscous fluids are to be handled, since high torque is required for operating at heavy *starting* loads (e.g. crushers/grinders).

Provide emergency power supply arrangements to keep the key equipments running/available for operational mode during failure of main grid supply.

11.5.1 Variable-Frequency Drives for Speed Control

These can reduce power consumption since it can make the equipments run at lower speed if there is less demand on the output. This is a better option than running the blowers/pumps at full speed and throttling the valves in the gas/fluid lines if requirement of fluid flow is less.

11.5.2 Consider for Use

- Copper bus bars are used instead of aluminium bus bars since copper has a higher melting point and better conductivity than aluminium. Optimise cost of bus bar by considering area of cross section required for same current to be carried.
- HT motors/energy-efficient motors instead of LT motors.
- LED lights instead of incandescent bulbs.
- Light reflecting colours and sunlight through transparent/translucent roof instead of electrical lights at the workplace.
- Better thermal insulation to minimise load on air conditioning and refrigeration systems.
- Chilled water (at +5 to +8 °C) instead of brine at (−5 to −8 °C) is preferable if the process unit can be run satisfactorily. Refrigeration plants consume less power in the previous case for operation.
- Electronic starters and chokes instead of conventional units.
- Graphite electrodes instead of water-cooled steel electrodes.

11.5.3 Electrolytic Processes (Examples-Manufacture of NaOH, Aluminium, Electrical Grade Copper, Gold Refining, Electrolytic MnO_2)

These processes consume a large amount of power. Even small voltage drops of 0.2 V in the circuits can waste lot of power since the currents are of the magnitude of hundreds or thousands of amperes. Hence, design, operation, and maintenance of electrolytic cells and their grouping (in series and parallel) shall be carefully looked into.

The following shall be optimised to reduce power consumption:

- Control the flow rate of electrolyte to each cell,
- Composition at inlet of the cell,
- Temperature of electrolyte in the cell,
- Distance between anode and cathode,
- Removal of gases from the vicinity of electrodes (gases are strong insulators),
- Voltage drop in contacts of bus bars with electrodes,
- Regularly clean contacts in bus bars and electrodes when heavy currents are passing,
- Geometry/shape of electrodes,
- Surface area of electrodes—optimum current density.

11.6 Heat Recovery and Cogeneration

Availability of energy:

Heat is evolved during many chemical reactions, and the products/process streams come out from the process reactors/units at considerably high temperatures. These streams are to be cooled in many plants for maintaining proper temperatures in downstream units for running the plant efficiently.

The possibility of recovery of thermal energy from such streams must be examined, and suitable heat recovery equipments, such as boilers, economisers, air preheaters, and recuperators, should be installed.

Certain exothermic reactions are carried out in reactors where cooling systems are to be arranged. Heat recovery can be considered from cooling jackets used for maintaining the reaction temperatures.

11.6.1 Cogeneration of Power Through Waste Heat Recovery Boilers and Economisers

Hot gas streams at 900–1000 °C are generated in sulphuric acid plants by burning sulphur. These gases are cooled by **waste heat recovery boiler** before further processing. High-pressure steam at pressures of 45 kg/cm^2 and temperatures of 410 °C is generated as a *by-product* which is supplied to a condensing turbine to produce power (which can be more than what is required to run the plant). This surplus can be exported to other industries.

Alternatively, this steam is used to run a backpressure turbine coupled to the main air blower (which needs considerable power). The exhaust steam from the turbine can be used for process heating, e.g., evaporation of alum or phosphoric acid solutions or for obtaining potable water from sea water (in certain locations) by multiple-effect evaporators, thus saving heat energy required for such processes. *This is cogeneration.*

11.6.2 Heat Recovery from Gases

Example
The gas streams come out from the converter passes of the sulphuric acid plant at sufficiently high temperatures (590–610 °C from the first pass) and need to be cooled further. Additional WHRB and economisers are installed in the plants which generate more steam and hot water.

The hot gases come out at 415–450 °C from the last pass and can be used along with an economiser for boiling 28–30 % oleum to produce pure SO_3 vapours.

Illustration
Please refer to the figure given below—hot process gases are passed through the WHRB for reducing their temperature before sending to downstream process units. High-pressure steam is generated by the boiler, and it is sent to the steam turbine which is connected to the air blower. The blower is also driven by an electric motor which gets power from grid supply.

The WHRB is sometimes bypassed to maintain temperatures in the downstream units. The steam generation gets *reduced* at this time, and the turbine cannot get full supply. However, the electric motor is always kept on, and hence, the air blower continues to run at a normal speed. The power drawn by the motor is reduced automatically to the extent of load shared by the steam turbine. Thus, the electrical consumption gets reduced even when steam generation is not steady.

The system works better when grid supply is available; the capacity (power rating) of the motor is chosen properly for the load (the air blower in this case), and the turbine shall not generate surplus power. In certain process industries, the turbine is a backpressure type, and the exhaust steam from the turbine can be used for process heating also.

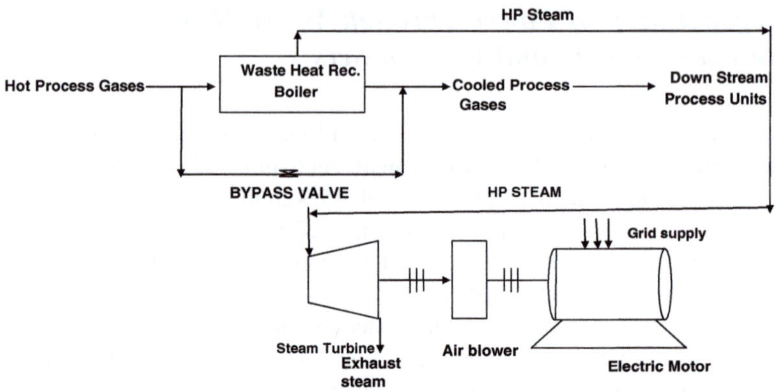

- Waste Heat Recovery Boiler is bypassed occasionally to maintain temperature in downstream units
- Steam generation is not steady
- Electric motor is always kept on.
- Supply of steam to the Steam Turbine is not constant.
- Draw of power from the grid reduces automatically as per steam supply to turbine.

Energy Recovery from Variable Flow Hot Gases

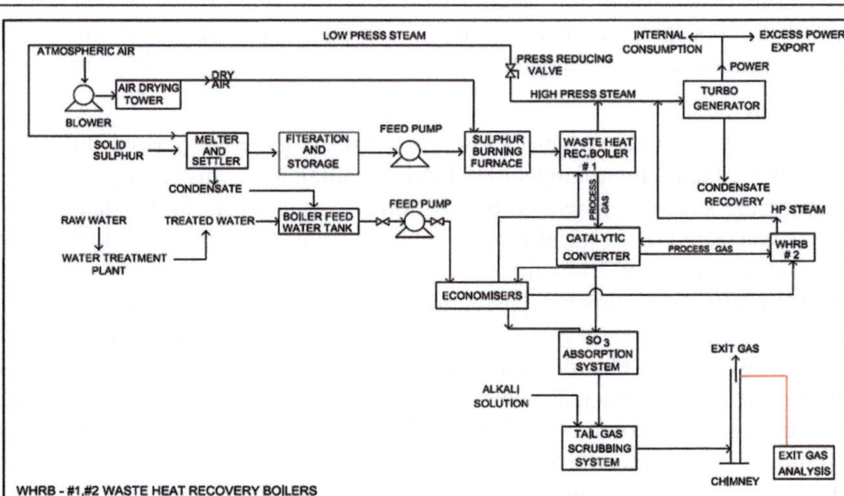

MANUFACTURE OF SULPHURIC ACID WITH HEAT RECOVERY

11.6.3 Heat Recovery from Hot Acid

- Sulphuric acid plants have circulating acid streams in the absorption towers. These streams come out of the towers as hot acid. Low-pressure steam can be generated by recovering heat from such streams of hot acid. Readers may contact Monsanto Enviro-Chem, USA, for more information since it is their technology.
- Hot water production for process use from hot acid by use of plate heat exchangers with special alloy construction (with dual cooling circuits) has been carried out by the author.

11.7 Efficient Use of Steam

The steam is generated by firing fuels such as coal, furnace oil, and natural gas or by WHR units in chemical plants. It is used for power generation and for process heating. Steam is thus a very convenient source of energy. It can transport large amounts of heat energy in the plant and deliver it efficiently due to very high coefficients of heat transfer. However, it is very necessary to use it most efficiently due to high cost of generation (from firing of fuels) or to minimise cost of production.

The important considerations for *increasing the system efficiency* while using steam shall be based on requirement of power to run the plant, possibility of exporting surplus power or to use the steam partly for generating power and partly use it as a heating medium.

11.7.1 Use for Power Generation

The selection of steam turbines and generators depends on the internal demand for power and heating duty. The surplus steam (available from exhaust of a steam turbine after power generation) can be used as follows: (*cogeneration system*)

- Sulphur melters and keeping sulphur feed lines and pumps warm,
- In multiple-effect evaporators (for use of vapours from first effect to heat the medium in second effect, for use of vapours from second effect to heat the medium in third effect, and so on)
- For drying wet solids.

11.7.2 Use as Heating Medium

The important considerations for using steam as heating medium are as follows:

- The required temperature to which the material is to be heated (for deciding the corresponding saturated steam pressure to be employed),
- the total quantity of material to be heated (or its flow rate per hour for calculating heat duty required to calculate steam requirement),
- whether a fast rate of heating is required (in case a batch process is being operated from a cold start every time), **OR**
  Is it required to take out the condensate as soon as it is produced if a continuous flow of material is to be heated for fast heating (the condensate will come out at saturation temperature of steam),
  OR
- Is the condensate to be allowed to cool down further (to recover heat from hot condensate also),
- Consider the preheating of incoming *cold material by use of hot condensate for better efficiency.*
- The vessel, heating jacket, and steam coils should be able to withstand the steam pressure (ensure proper mechanical design of these),
- Recovery of heat value of low-pressure steam: by mechanical vapour compressors to increase its pressure by compression, reheat it, and again use it.

11.7.3 Selection of Steam Traps

Steam is used in many chemical industries for the following:

- Heating of reactors (manufacture of organic chemicals) and heating of storage tanks by inner coils or external jackets (holding viscous oils),
- To keep the pipes warm (to ensure they can be pumped in cold weather also) by having another larger diameter pipe over it and passing steam in the annular space, i.e. by jacketing, or by providing a coil just touching the external surface of the tanks instead of installing inside, or by laying a parallel pipeline carrying steam touching the pipe carrying process fluid—if any ingress of steam through a leak can cause a mishap—this is known as steam tracing and is a safe method for such applications.
- Steam is also used to melt sulphur in underground melters, to keep jacketed pipes carrying molten sulphur hot.

Steam can supply a large amount of heat over long distances (as it has a large latent heat and does not need a pump).

Suitable steam traps shall be provided instead of providing valves for automatic removal of the condensate.

11.7.4 Steam Traps

Steam traps must be able to:

- Remove air and the **non**-condensable gases since these are poor conductors of heat and can reduce the heat transfer coefficients in the system. They can interfere with contact of steam with heat transfer surface.
- Remove the condensate formed.
- Prevent escape of live steam (*it is necessary to minimise the loss of live steam since it is costly to generate*).

11.7.5 For Considerations by the Plant Engineers

Plant engineers should consider the following for selecting a suitable steam trap:

- Process requirement—melting a fixed amount of solid or evaporating a fixed amount of liquid per batch. The steam requirement reduces after the batch operation is over.
- To maintain a particular temperature in the process vessel to keep it hot—this is to compensate for the heat losses from the body of the vessel. *Steam consumption is less all the time.*
- Continuous melting or evaporation process—there can be a higher steam demand all the time the operation is going on.
- If other gases or non-condensables are also present in the steam—e.g. in case of later effects of a multiple-effect evaporator, **the trap should be able to remove gases**.
- If it is desired to recover some more heat from the hot condensate also—the condensate is to be allowed to cool down further instead of immediate removal as soon as it is formed.
- Expected backpressure from condensate recovery pipeline if it is too long or *in case the condensate is at high pressure—**when being discharged**—it can flash into live steam. This may be lost to the atmosphere.*
- Steam pressure used for heating and corresponding saturation temperature shall also be considered.

11.7.6 Some Typical Steam Traps

Some typical steam traps are as follows:

Mechanical traps (inverted bucket type)
They have a float that rises and falls in relation to condensate level in the body of the trap. This type of trap *can drain out condensate continuously.*

Thermostatic steam traps

- **Bimetallic trap**: An oil-filled element expands if it comes in contact with live steam and it closes the condensate exit valve. It may be possible to adjust the discharge temperature of the condensate from the trap. Hence, they can hold back condensate till it cools sufficiently. The condensate is not removed immediately as soon as it is formed. *The heat transfer coefficient in the system being heated is less due to this.*
  This makes the thermostatic trap suitable to remove large quantities of air and cold condensate at the start-up condition, especially for batch-wise processes.
- **Thermodynamic (TD) traps**: As steam enters, static pressure above the disc pushes it against the valve seat. It does not allow live steam to escape. As the steam starts to condense, the pressure against the disc reduces and opens the valve, allowing the condensate to go out. *It is thus able to remove condensate fast.*

Chapter 12
Methods for Heating

12.1 Introduction

Many chemical plants need properly dried, calcined or preheated input (raw) materials before feeding into the reactors and process units for production of chemicals. It is necessary to carefully consider the various issues for **heating arrangements for a chemical plant** from safety, convenience, and cost points of view.

12.2 Heating Load and Temperatures Required

- Thermal and physical properties of the material to be heated—specific heat, latent heat, melting and boiling points, thermal conductivity, flash point, viscosity, density, and evolution of dissolved gases/vapours generated during heating;
- Quantity of material in the equipment (for batch operation);
- Flow rate into the unit and going out from the unit (for continuous operation);
- Initial and final temperatures of material to be heated;
- Initial and final temperatures of heating medium.

12.3 Broad Selection of the Heating Method

(i) **System features**

- Construction of equipment,
- Properties of material to be heated,

K.R. Golwalkar, *Production Management of Chemical Industries*,
DOI 10.1007/978-3-319-28253-4_12

- Sources of heat to be considered from,
- Reliability required for operation,
- Safety and ease of operating the system,
- Space available for the equipment,
- Budgetary cost of system with accessories required.

(ii) **Operational considerations**

- Accuracy in temperature control required,
- Whether heating is to be done continuously or intermittently,
- Safety and ease of handling the heating medium/fuel and operation of the system (including storage arrangements at site),
- Time required for reaching operating temperature from cold start,
- Time required to cool down after the feeding of fuel has been stopped,
- Cost of operation and maintenance.

12.4 Factors to Be Considered for Selection of Heating Method

12.4.1 Types of Equipment

- Reactors: stirred tank with internal coils or external jacket, catalytic reactor with fixed bed/fluidised bed;
- Evaporators: single-effect (open pan type); multiple-effect evaporators of various types (readers may refer to **Chemical Engineers' Handbook**);
- Tray dryers—fixed bed type for drying of wet solids;
- Rotary dryer (jacketed) or direct heating of material;
- Rotary kilns;
- Melters and dissolvers for solids;
- Shell and tube type heat exchangers;
- Steam distillation units for heat-sensitive materials;
- Furnaces of various types.

12.4.2 Types of Processes

Types of processes—evaporation at reduced pressures, heating or cooling of fluids; calcinations to remove volatiles, drying of wet products to remove moisture, evaporation of heat-sensitive products, and batchwise concentration or continuous evaporation at ambient pressures.

12.4.3 Space Availability

Space available for installation of accessories required for the heating method chosen for the equipments.

12.4.4 Operating Temperatures

Operating temperatures—high (>500 °C), middle range (250–500 °C); lower side (60–250 °C).

12.4.5 Temperature Control Required

Temperature control required—accurate (within 4–5 °C of set value), moderately acceptable range (10–15 °C), within a larger variation (20–30 °C).

12.4.6 Purity of Material to Be Maintained

Purity of material to be maintained—no charring of products, no deposits of ash, dust particles or scales.

12.4.7 Availability of the Heating Medium at Site

Availability of the heating medium at site—liquefied petroleum gas, natural gas, coal, electricity, process gases, and steam.

12.5 Construction of the Equipment

- Reactors: (can be a catalytic reactor with fixed bed/fluidised bed)
- Jacketed from outside: both the equipment shell and the jacket are subjected to the heating medium. The shell is also in contact with the reaction medium from inside;
- Equipment with internal coils: The equipment shell is not subjected to the heating medium but is in contact with the reaction medium from inside. The

coils are in contact with heating medium from inside and the reaction medium from outside.

- Equipment is lined inside with refractory bricks for protection of the shell: The shell is not subjected to the heating medium as well as the reaction medium. This type of equipment cannot have heating jackets due to poor thermal conductivity of the bricks. It has electrical heating by electrodes installed inside or has induction heating method. It can also have internal heating coils of special MOC which can resist the material to be heated.
- Evaporators:
  Single-effect or open pan type evaporator—open to atmosphere, heated from below by flue gases or burning of fuel like coal;
  Multiple-effect evaporator of various types—for sugar plant, concentration of sodium sulphate solution in rayon plants.
- Tray dryers—fixed bed type for drying of wet solids like dyes for paper mills.
- Rotary dryer (jacketed) or direct heating of material—drying of wet sodium sulphate.
- Rotary kilns—cement production, incineration of wastes.
- (i) Melters-(underground pits with steam coils for sulphur in sulphuric acid plants);
  (ii) Dissolvers for solids—for dosing chemicals in waste water treatment plants.
- Shell and tube type heat exchangers—boiling of 28 % oleum for generating SO_3 vapours (heating mediums used are steam or hot process gases).
- Steam distillation units for heat-sensitive materials—production of essential oils from aromatic plants;
- Furnaces of various types—MnO_2 reduction furnaces, CS_2 manufacture furnace.

12.6 Minimising the Loss of Heat from Process Units

- To reduce loss of heat from the process units so that fuel consumption can be reduced,
- The required temperature level is achieved faster,
- Quicker resumption of production every time the unit is heated,
- Minimising heat loss during running,
- To minimise fall of inside temperature by retaining heat inside *whenever the operations are stopped due to any reason.*

12.7 Methods for Conservation of Heat/Reducing Requirement of Heat

- Better refractory brick-lining and insulating ceramic papers,
- External insulation of optimum thickness and aluminium cladding,
- Shining surface paint/surface finish,
- Install process unit in shed (protect from rain, snow fall, and strong winds),
- Avoid frequent start and stop,
- Operate at lower temperature if possible.
- Using hot concentrated solution from process unit (*evaporator*) exit to preheat incoming dilute solution is an efficient method of heat recovery;
- Use of hot water blowdown from boiler to preheat incoming cold feed water;
- Put blinds in gas ducts at inlet and exit of process units to conserve heat during plant shutdowns.

12.8 Various Methods for Heating

12.8.1 Process Heating by Steam

High-pressure steam as a heating medium is one of the most common option due to the high heat transfer coefficient of condensing steam and the large amount of latent heat (around 510–530 kCal/kg, depending on the pressure) available per kg of steam condensed. No pump or blower is required for the transfer of steam since it can be transported to long distances by virtue of its pressure. It is thus very useful for heating in a chemical plant. It is a non-polluting, with fast heating, and a clean medium.

It can be used for cogeneration of power also.

Steam pressure of 10 kg/cm^2 and more is be required for achieving 150 °C and higher temperatures. This needs specially designed, fabricated, and tested process vessels and pipelines, besides requiring approval of statutory authorities.

It can be generated by coal-fired boilers/or will be available from waste heat recovery boilers (WHRBs) which are heated by hot process gases, hot exit gases from the plant.

12.8.2 Considerations for Buying a Coal-Fired Boiler

- Heat duty load for the process (heating a reactor jacket; evaporating a solution),
- Required temperature of process unit and corresponding pressure of steam,
- Amount of steam required for process heating,

- Type of boiler required—smoke tube (preferable for process heating) or water tube if specially required by the plant management,
- Manufacturer of boiler shall guarantee the steam generation at required conditions, and the coal/fuel consumption. The quality of feed water and coal required shall also be specified by manufacturer.
- Details of coal storage at boiler inlet and feeding arrangement. This may include coal conveyor, crusher, elevator, screw feeder along with their motor HP;
- Efficiency of combustion (loss of unburnt coal through grates not to exceed 1.0 %);
- Power consumption and motor details for ID fan, oil feed pump (for start only), etc., shall also be given;
- Standard fittings, mountings (safety valves, level indicators, feed check valve, non-return valve, vent, blowdown, mainsteam stop valve, etc.) provided;
- Boiler feed water pumps with motors duly assembled on common base plate;
- Automatic blowdown for control of concentration of dissolved solids inside boiler;
- Details of overall dimensions, weight of boiler when empty and when full;
- Civil foundation details;
- Approval of all pressure parts from statutory authorities;
- Erection and commissioning assistance;
- Instrumentation: level controller, pressure gauge, conductivity meter for boiler feed water (optional), etc.;
- Safety interlocks provided (high steam pressure and low water level trip);
- Power consumption per MT of steam generated at specified pressure and temperature.

Recommended boiler and feed water quality

Feed water to boiler limits

Pressure	Up to 20 kg/cm^2
Dissolved oxygen	Less than 2 ppm
Total silica (max) ppm	1
pH	8.8–9.2
Hardness, ppm	1

Boiler water (inside the boiler) limits

Pressure	Up to 20 kg/cm^2
TDS, ppm	3000–3500
Specific electrical conductivity at 25 °C	1000
Silica, ppm	25

12.8.3 Electrical Heating

It is easy to operate, has fine temperature control, and is clean, but it is very costly. It needs accessories as transformer, circuit breakers, and cables/bus bars of adequate rating for high currents, instruments such as ammeter, voltmeter, KW and KWH meters, and temperature controller thermostats (to switch on and switch off power automatically as per need), or by adjusting the power input by controlling the voltage across the electrodes.

Electric furnaces are equipped with graphite electrodes generally as these are capable of withstanding very high temperatures, and good conductors of electricity. The furnaces are provided with inner lining of refractory and insulating bricks to conserve the heat and to protect the outer shell. Heating takes place by electric arc between the electrodes or due to resistance of the material being heated. The electrodes need good supports and arrangement to push in new pieces for replacing the old ones (which get consumed slowly over time).

The efficiency is high since the heat is generated inside the furnace instead of heat being supplied from outside in case of fuel-oil-fired furnaces.

There is possibility of mishap when electricity is used for heating (due to short circuit or malfunctioning of thermostat).

12.8.4 Heating Directly by Flue Gases

Heating directly by flue gases can achieve high temperatures. However, the heat transfer coefficients on gas side are low and can get affected by deposits of soot, ash, etc. A larger heat transfer area therefore becomes necessary. The heating by direct use of flue gases is not suitable for inflammable liquids/in areas handling explosive dust. It can char heat-sensitive materials.

12.8.5 Heating by Hot Process Gases

Heating by hot process gas is an attractive energy-saving option. This needs careful study of the process operating conditions for the heating loads, and heat available from the hot process gases. Hence, consider the composition, specific heats, temperature of incoming and outgoing gases, and pressure at which the gases will be available and pressure drop permissible on gas side, *e.g., use of exit gases from economiser of a sulphuric acid plant for boiling 28 % oleum to generate* SO_3 *vapours which are then condensed to get liquid* SO_3.

Limitation—There could be a process requirement to heat the downstream units by taking the hot gases directly to them and **not** passing the gases through the heating unit for some time (this can reduce the availability of heat). *Hence, it may not be always possible to use this option.*

12.9 Factors to Be Considered for Selection of Fuel

- Cost of fuel including basic cost plus transport charges, facilities and manpower required for unloading and storing, insurance premiums for storage, etc.
- Net calorific value (NCV): heat available after complete combustion of the fuel *—this shall not be confused with gross calorific value (GCV) while comparing different fuels.*
- Cost of heat available per unit of local currency shall be considered, because certain fuels may have more NCV but could be very costly. This will also depend on the efficiency of combustion system being considered. Examples are as follows:

  (i) *Cost of coal and NCV are Indian Rs. 4/- per kg and 4000 kcal/kg, respectively. Efficiency of combustion is 80 %.*
  Hence, 4000 × 0.8 = 3200 kcal will be available at a cost of Indian Rs. 4/-. That is 800 kcal/Indian Rs. spent on fuel.
  (ii) *Cost of furnace oil and NCV are Indian Rs. 50/- per kg and 10,000 kcal/kg, respectively. Efficiency of combustion is 85 %.*
  Hence, 10,000 × 0.85 = 8500 kcal will be available at a cost of Indian Rs. 50/-i.e. 170 kcal/Indian Rs.

- Density of fuel kg/litre for liquids and kg/Nm^3 for gas—this should be known to calculate the volumes of storage tanks for liquids and then design them. Density of fuel gases shall be known at the pressure at which they are stored.
- In case of solids, the bulk density and angle of repose of the pile of solids shall be known for calculating the volume of shed.
- Physical properties like pour point, flash point, boiling point; vapour pressure at ambient temperatures (high during summer in certain countries).
- Higher amounts of moisture, sulphur, and ash % in fuels (these are more in coal, petroleum coke as compared to liquid and gaseous fuels) can cause deposits/corrosion of heat transfer surfaces. *They need special pollution control equipments like desulphurising scrubbers, electrostatic precipitators for flue gases.*

  - Viscosity at low temperatures—it may be difficult to pump these oils without heating in certain countries (pour point—to be considered).
  - Flash point—to be considered from fire hazard point of view.
  - Since liquid fuels and natural gas have better atomisation and stable flame characteristics, they need only a small amount of excess air for proper

combustion. The volume of flue gases generated is therefore less as compared to the volume of flue gases produced when solid fuels are used.
- Higher temperature and better efficiency can be achieved as a result with the former.
- Larger volumes of flue gases carry away more amount of usable heat in case of solid fuels.
- **The design of furnace** (location of air injection nozzles, refractory linings and walls, fuel/air ratio controllers), heating equipment, and burners also depends on the type of fuel and its properties. It can influence the efficiency of combustion, the accessories required for using the fuel (*coal feeding and ash handling system to be compared with oil pumps used for feeding furnace oil*).

The particular construction of the furnace can affect the time required to heat up from cold start to reach operating temperature, the loss of unburnt fuel, the production held up during the heating period (as well as the heat wasted during the cooling time when the operations are to be stopped for taking out products or ash; and for maintenance of furnace or accessories for combustion of fuel). Hence, the amount of *heat actually* available for heating purpose is to be considered while choosing the fuel.

- Arrangements for fire fighting and safe handling with gas detectors, earth connections for storage tank, lightening arresters and dyke walls around storage tanks shall be provided when liquefied gaseous fuels or fuel oils are stored. **These should be as directed by statutory inspection authorities**. *It is dangerous to store too much liquid/gaseous fuels.*
  - Site infrastructure and facilities required, e.g. unloading, storing, and transferring to points of use in production plant.
- Special facilities/accessories required for using the particular fuel—*given below.*

12.10 Accessories/Site Facilities Required for Using Various Fuels

12.10.1 Gaseous Fuels

- Natural gas from oil wells,
- Liquefied petroleum gases—from petroleum refineries,
- Producer gas,
- Hot process gases which are to be cooled and are available in process plants,
- Flue gases after combustion of fuels,
- Hot air—from air preheaters, from hot air generators (using various fuels, electrical power, steam).

12.10.1.1 Features of Gaseous Fuels

- They do not contaminate the heat transfer surfaces/the product with ash deposits.
- Use of such fuels can be a very clean and efficient option.
- Reliable external source is preferable, since storage of large quantity of gas in the premises can be dangerous.
- *Convenient when they are obtained from a direct pressurised gas line connection with an external gas grid, or directly available from oil fields at certain locations.*
- Temperature control is accurate and easy. Combustion of gaseous fuel generally does not cause pollution unless it contains sulphurous compounds. *Tail gas scrubbing system shall be installed when such gases are used.*
- Igniting the gas in a furnace and completion of combustion is easy as proportions of gas and air can be controlled conveniently (and automatically).
- Gas firing can be quickly started, the flame can be stable, and high temperatures can be achieved due to easier control of excess air.

12.10.1.2 Accessories Required When Gaseous Fuels Are Used

- Flow control valves, pressure regulator, and gas flow measurement meters shall be fitted on incoming gas lines. These shall be of statutorily approved make only and meet all safety standards as directed by inspection authorities.

12.10.1.3 Necessary Safety Precautions

- All pipelines, valves (including shut off and non-return valves) fittings shall be as per international standards (pressure tested and corrosion resistant). These shall be tested regularly and faulty parts replaced by new ones immediately.
- Layout of the gas pipelines must avoid proximity to electrical cables, furnaces, and process ducts operating at high temperatures, areas where naked flames could be present.
- Automatic condensate drainage from gas lines shall be provided.
- Detectors for inflammable gas leak and warning devices must be provided.
- Vehicle parking should not be permitted in the vicinity of liquefied petroleum gases storage tanks due to chance of fire caused by sparks flying from exhaust pipes of the vehicles. All vehicles in the area must have spark arresters on their exhaust pipes.
- All nearby electrical equipments and fittings shall be flame proof.
- Control valves in gas feed lines shall be photocell actuated and shall stop gas flow if the flame is extinguished due to any reason.

- The gas and air flows shall also be regulated automatically as per temperature setting of the equipment/material being heated. CO_2 % in exit gases may also be monitored and used for air flow control.
- Forced draught fans and tail gas scrubbers (if the gases contain H_2S or other sulphurous compounds) are to be provided as per need. The gas supply shall be automatically cut off if the air supply pressure to furnace gets reduced below set point due to any reason. This is to prevent explosions.
- A flare tower shall be available with a continuous pilot flame to release any excess pressure of fuel gas coming from external gas grid. This can happen when there is no demand in the process plant and the shut off valve is closed.

12.10.2 Liquid Fuels

Common liquid fuels are high-speed diesel (for special application only), kerosene, furnace oils, and light diesel oils.

12.10.2.1 Features of Liquid Fuels

- They do not generally contaminate the heat transfer surfaces/the product with ash deposits.
- Use of such fuels can be a very clean and efficient option.
- Feeding at steady rates is convenient by metering pumps. The consumption can be monitored easily.
- Storage of large quantity of liquid fuels in the premises can be dangerous. A reliable source for regular replenishment shall be arranged to reduce inventory.
- Temperature control is accurate and easy.
- Combustion of these fuels generally does not cause pollution unless they contain sulphurous compounds. *Tail gas scrubbing system shall be installed in such cases.*

12.10.2.2 Arrangements and Accessories Required for Liquid Fuels

- Receiving fuel supplies by special tankers, weighbridge at site, and unloading facilities for filling into main storage tanks.
- Main storage tanks shall have level indicators, cooling arrangements by spray of water in hot weather, and safety vents.
- Heater at outlet of tank with thermostatic temperature control (electrical heater) or a steam pressure regulator (if steam is used for heating of oil), a temperature

gauge, and provision of strainers and suitable transfer pumps for oils which have high viscosity.

- Dyke wall enclosure to collect any spillage/leak.
- Statutorily approved fire fighting arrangement, lightening arresters.
- The tanks shall be placed on I-beams which are fixed on civil foundations. The tank bottom plates shall have anti-corrosive paint.
- Pipelines valves and gaskets suitable for high pressure and temperatures, external insulation.

  – Photoelectric cell flame monitor for furnace.

- Oil transfer pumps shall have mechanical seals and flame proof motors. Gear pump with in-built safety devices preferred for pumping of viscous oil, and filters at oil burner inlet.

  – Smaller capacity (for 4-h operation) day tanks shall be installed in the process plant near the oil firing burners. These tanks shall be filled up every 2–3 h.
  – Oil transfer lines shall have good external insulation, covers on all flanged joints, and shall have a route away from electrical cables or high voltage units.
  – The furnace shall have metering fuel pump, a pressure regulator, internal safety valves, oil filter at appropriate place in oil line, oil burners with refractory lining and flow controllers.
  – An oil flow meter with totaliser shall be provided to monitor total quantity of oil consumed.
  – Spare atomising nozzles of different sizes shall be available with plant engineers.
  – Air blower and compressor (positive displacement type) are preferable with silencer-cum-filter, dampers for primary (atomising) air, primary and secondary air piping, manometer, pressure regulators for air supply at burner inlet. The positive displacement-type combustion air blowers shall also be able to atomise the oil or separate compressors for atomisation of oil shall be provided.
  – Indicator/controller of CO_2 content in exit gas to ensure proper combustion.
  – Presence of soot particles shall be monitored in exit gases. Black-coloured exit smoke is indication of incomplete combustion (may be due to improper atomisation or insufficient combustion air or both).
  – Control valve in oil feeding line shall be actuated by photoelectric cell (to stop oil pump if flame goes out).
  – Oil pump motor shall trip if combustion air pressure at furnace inlet is too less.

12.10.3 Use of Solid Fuels

Some common solid fuels are coal, pet coke, industrial combustible waste.

12.10.3.1 Features of Solid Fuels

- Biomass and agro waste Solid fuel firing system generally requires more manpower than gaseous/liquid feeding system for unloading, storing, and feeding the solid fuels.
- Fine temperature control is difficult to achieve.
- Chances of loss of unburnt fuel if the pulveriser is not working properly.
- Ash removal system is necessary as well as arrangements for its disposal.
  - *It also needs more time to start firing and to reach required temperature and to stop and cool down.*
  - *Regular removal of the ash from the furnace is required which may pose disposal problems besides needing equipment for removal and handling.* Loss of heat transfer efficiency may occur due to ash deposit on heating surface when coal is fired. There is also the problem of ash disposal. *These problems do not arise when oil firing is used.*
- Pollution control devices such as electrostatic precipitator and flue gas desulphuriser are required if ash and sulphur content are more in the solid fuel (e.g. some varieties of coal, and petroleum coke). *These equipments are costly.*
- *In case of liquid or gaseous fuels, quick start and stop is generally possible; and they have less ash content. The coal-firing unit cannot have quick start to generate steam as soon as it is required. It also takes a long time to cool down. Hence, there is considerable loss of heat during starting and stopping.*
- Due to above reasons, coal firing shall be used only when the plant is to be run continuously for long durations.

12.10.3.2 Site Infrastructure Required

- Sufficiently large storage shed to protect from rains, dust.
- Unloading, weighing, and transfer arrangements.
- Feeding system to furnace.
- Storage bins and feeding hoppers.
- A pulveriser is required for a coal-firing system along with other accessories such as screen for preventing entry of large lumps; and recycle arrangement for oversize material; conveying system (bucket/belt conveyors with arrangement for weighing the coal transported), storage bunker, and a screw feeder for

feeding into the furnace. This may result in considerable power consumption (*shredder and dryer are required for biomass*).

- Air blowers (FD fan and ID fan), CO_2 % analyzer for exit gases, air preheater, chimney, arrangement for ash removal, its collection, and disposal.
- Auxiliary firing arrangement with oil firing may be required in case of solid fuels for starting combustion and in case of disturbed flame due to any reason. This is to maintain flame stability if coal is wet or not pulverised properly.
- Rice husk, bagasse, saw dust, dry grass, agrowaste have low bulk density cost, and hence, cost of transportation is high (not much material can be loaded in one truck load as compared to coal). Their availability is also seasonal, and hence, using in industry is comparatively less. They also need more storage space.

12.10.4 Industrial Waste

Industrial waste (oily sludge from tanks, paint shop waste, sludge from cleaning of some process reactors).

Extensive laboratory analysis and statutory permission are necessary to decide whether these can be used as industrial fuel or sent to destruction by incineration. The flue gases produced when this type of fuel is burnt may contain considerable pollutants. They need air pollution control system having venturi scrubbers, packed towers, electro static precipitators, induced draught fans and a tall chimney.

12.11 Combustion Air for Firing Systems

- Combustion air is generally consisting of primary air (for atomisation of liquid fuels), secondary air for combustion and tertiary air for ensuring completion of combustion of any remaining oil droplets/fuel particles and for protecting furnace refractory lining. These are controlled by suitable control dampers in the air lines.
- Preheated air can be obtained by heat recovery from exit gases. It can increase furnace temperature, can reduce fuel consumption (by minimising loss of unburnt fuel droplets), and will be found useful for completing the combustion.
- Excessive air can decrease furnace temperature, or the flame can get extinguished.

Since liquid fuels and natural gas have better atomisation and stable flame characteristics, they need only a small amount of excess air for proper combustion. The volume of flue gases generated is therefore less as compared to the volume of flue gases produced when solid fuels are used.

- Larger volumes of flue gases carry away more amount of usable heat in case of solid fuels. Hence, excess air volume shall be controlled by monitoring CO_2 and O_2 % in exit gases.
- Too less an amount of air can cause explosive mixture inside furnace or deposits of unburnt fuel particles on downstream equipments.
- Positive displacement-type blowers will be found useful (*if available in the required capacity at a reasonable cost*) for atomisation and for supplying required air.
- Forced draught (FD) and induced draught (ID) fans: It may be considered for handling large gas volumes of the flue gases and discharging through a tall chimney. The ID fans shall be fabricated from corrosion- and erosion-resistant MOC.

12.12 Safety Devices to Be Regularly Checked by Senior Engineers

- There are chances of explosion or unburnt oil drops going to downstream units if the oil pump continues to supply oil even when air supply pressure at inlet of furnace is too low due to any reason. Electrical interlock of oil pump with manometer indicating air pressure shall be made. The oil pump must stop if air pressure is too low due to any reason (see Fig. 6.5).
- Similarly, supply of gaseous fuel shall stop if air pressure is too low.
- Provision of photoelectric flame monitor with warning alarms and tripping system for fuel supply shall also be made if flame is extinguished due to any reason.
- Combustion chamber shall have appropriate refractory lining (insulating and refractory materials), **safety vents/rupture discs,** thermocouples for temperature monitoring, external insulation, and a shed for protection against rain.

12.13 Heat Transfer Oils (Thermic Fluids)

Steam is one of the most common options for heating. However, a steam pressure of 10 kg/cm^2 and more will be required for achieving 150 °C and higher temperatures. This needs specially designed and fabricated vessels (heating jackets and coils), steam pipelines, and valves, with the approval of Pressure Vessel and Steam Boiler Inspection Authorities in the country. This is not necessary when heat transfer oils are used for heating even up to 270 °C as the system can operate at atmospheric pressure.

Heat transfer oil (HTO) has high boiling point and thermal stability at higher temperature. The HTO is heated by burning of coal, furnace oil, or light diesel oil, and it is circulated by a pump through the jackets or heating coils of the process vessels. The process vessel and jacket need not be operated at high pressure as in the case of heating by high-pressure steam.

Obtain a sample of the HTO for testing for the thermal stability (it shall not form any thick deposits in the tanks or pipes during the high-temperature operation). Also check the vapour pressure data for HTO oils while heating the oil beyond 200 °C.

Caution: There are chances of fire due to the circulating high-temperature oil, and hence, the motors, electrical fittings, and lighting must be flameproof. **The pipelines, valves, gaskets, and pumps (with mechanical shaft seals) must be selected for high-temperature and high-pressure service and shall be tested as per statutory requirements.**

Since the heat transfer coefficient is much lower with HTO as compared to that of condensing steam, the HTO fluid will have to be heated about 20–30 °C above the required temperature for proper transfer of thermal energy to the material in the vessel.

Accessories required for use of HTO system:

- A high-alumina (not less than 45 % Al_2O_3) refractory-lined firing chamber, furnace oil, or light diesel oil fuel tank and firing system, with fuel oil tank, air blower, and chimney for flue gases.
- HTO circulation system (circulation tank, pump, expansion tank, pipelines and valves; instrumentation and alarms) and suitable insulation and cladding on the hot oil tanks, piping, and valves.
- Total volume of circulating HTO in the system shall be checked carefully since it can expand on heating by 7–8 %. The volume of the expansion tank shall be sufficient for this. Vents shall be available on the expansion tank.

- **Parameters to be monitored and controlled**:

  i. Flame stability (through photo cell) if liquid fuel is used for heating the thermic fluid;
  ii. Temperatures of hot thermic oil at the circulating pump discharge, at inlet and outlet of the process vessel;
  iii. Level in expansion tank and any overflow from vent line.

- **Instrumentation and safety devices** for high-temperature cut-out for stopping oil firing in case of high HTO temperature; furnace oil supply cut-out if the burner flame gets extinguished or low pressure in air supply lines (due to blower tripping, closing of valve in air lines, etc.).
- Audio–visual alarms (siren and warning lamps) shall be available.
- Circulation tank for HTO shall be away from firing chamber. Any leak or spillage of HTO shall not fall into the firing chamber. All flanged joints shall have covers.

Chapter 13
Inventory Control

13.1 Introduction

Chemical industries need various types of items for their safe, pollution-free operation. The process units and machineries need reliable spare parts for the smooth running. The wear and tear of various equipments is comparatively more in chemical industries than other industries due to corrosive nature of many chemicals and presence of fumes, dust, and gases.

The various machineries for smooth conveying of solid materials are belt, screw and bucket conveyors, hoists, overhead cranes, and fluid transfer equipments such as pumps of various types, air blowers, and compressors. Many different types of processing reactor, filters, agitated vessels, condensers, gas absorbers, etc., are also required.

They need spares such as belt fasteners, idler pulleys, buckets and chains, bearings and shafts, impellers, filter media, tower internals, heat transfer tubes, drive motors, etc. It is very important to always maintain adequate stock of all important components and spare parts because unsafe conditions can occur, environmental pollution may result, and production can get interrupted for want of these spares, e.g,

- Non-provision of rupture discs on pressure vessels will make operation dangerous;
- Non-replacement of leaking mechanical seals for shafts of pumps handling toxic chemicals can cause environmental pollution;
- Continuing with old corroded parts of flow control valves can upset the process;
- Not adding stabilisers for preventing solidification of SO_3 may make it very difficult to handle it.

Hence, it is necessary to correctly estimate the rate of consumption of various items at maximum production capacity *instead of the rated* production capacity of the plant. Some typical consumables and spares are catalysts, gaskets, special pipes,

K.R. Golwalkar, *Production Management of Chemical Industries*,
DOI 10.1007/978-3-319-28253-4_13

baffles and agitator blades, effluent treatment plant and water treatment plant, chemicals, stabilisers, refractory bricks and cements, demisters in absorption tower, thermowells, gland packing materials for pump shafts, personal safety devices, laboratory equipments and reagents, etc.

This can be done by properly recording the replacements of spares carried out and other items consumed over 6 months to 1 year. The figures for stock levels could be higher than the figures recommended by vendors and can increase the cost of inventory. However, it is better to be on the safer side.

A list of all items and spares (*based on the above recommendations and own records*) shall be prepared. This shall include the major spares as well as additives, stabilisers, and property modifiers which are generally required in small amounts *but are costly* and hence their stock levels shall be carefully estimated. The list can also include various types of containers and packing items such as carboys, drums, and boxes.

13.2 Record of Past Consumptions

A study of the record of spares replaced in last few months/year shall be done to know which ones are being procured more frequently. The reasons for this shall be investigated. *Formats have been given in Appendix for keeping such records required for this study.*

- Some of the equipment might have been operated beyond their maximum limits of temperatures (for heat exchangers), maximum operating speeds (for agitated reactors), or maximum permissible delivery capacities or backpressures (in case of pumps).
- Attempts shall be made in consultation with design engineers, consultants, shop floor engineers, and experienced workers to reduce the rates of consumption of as many items as possible. Initially, address those items which have high individual cost or are very critical for operating and maintaining the plant. These could then be followed by items which are less costly or are less important.

13.3 Use of Guide Words

13.3.1 Simplify

Plant engineers shall try to simplify the equipment/process operating conditions.

- Try to operate at lower pressures and temperatures *without disturbing* production rate, product quality, and energy efficiency.
- Provide calibrated day tanks, rotameters, and metering pumps for controlled feed.

- Provide easily removable baffles and agitators with removable blades of different designs and sizes.
- Process operating personnel can also help further by operating the equipments using less concentrated acids, by loading the machinery more carefully (*see below*), or slower speeds of agitators.
- Typically, rate of charging of raw material bags into the reactor can be controlled. Instead of adding all the bags together (which can suddenly overload the agitator, its gear box, and the motor), the bags may be slowly put in one by one if manually charged in the reactor. A screw feeder with controlled speed can also be considered for feeding the raw material into the reactor.
- Use filtered input solutions (for removal of erosive hard particulate matters) instead of charging in solid powders directly into the reactors.

Case study

Concentrated acid was being added through a 50-mm-diameter pipe from its open end in the centre of a reactor. Specific gravity of the acid was higher than the reaction medium, and hence, it was going down and reaching the agitator blades of fibreglass-reinforced plastic. The blades were getting attacked by the strong acid and getting corroded. Hence, a *long pipe with a number of small holes* was used for the addition. This could uniformly and slowly add the acid along a larger cross section in the reactor and hence got mixed properly in the reaction medium. The attack on blades of the agitator got reduced as a result.

***Please see Figure at Art 8.4 in Chap.* 8.**

Operating personnel shall carefully monitor any abnormal noise, higher load on drive motors, and any vibrations and attend any such abnormality immediately.

13.3.2 Substitute

- Substitute strong acids by less corrosive reaction medium; use properly pulverised and screened materials instead of bigger lumps of raw materials; or use solutions already prepared in advance instead of charging solids into the reactors.
- Use hot treated water or hot condensate for reaction medium instead of using cold water and heating the jacket by steam.
- Substitute carbon steel shafts by stainless steel or by lead-bonded steel in corrosive conditions for longer life.
- Substitute ordinary gland packing at shafts by water-cooled gland packing. This will reduce the escape of hot vapours (which can attack the gear box/drive motor above).
- Use toughened glass tubes instead of ordinary glass tubes for level indicators.
- Use Poly Tetra Flouro Ethylene (PTFE) gaskets instead of compressed asbestos gaskets for pipeline carrying oleums/very corrosive acids.

13.3.3 Strengthen

- Make the equipment/components/parts stronger;
- Provide with special tubes (A-179 grade) for high-temperature service and refractory ferrules at gas inlet of the tubes of heat exchangers;
- Choose larger diameters for the shafts for agitators and flow control valves for more strength;
- Provide thicker flaps for butterfly valves in gas lines;
- Provide external reinforcement ribs on the vessels for increasing their life.

13.3.4 Safety Features

- Prevent possibility of overloading of hoists by providing warning alarms and tripping of drive motor if there is excess weight.
- Prevent possibility of over travel (vertical as well as horizontal) of hoists and travelling cranes by providing limit switches for warning by alarms and by electrical tripping of the drive motors.
- Provide lock nuts on foundation bolts to prevent their loosening (which can cause vibrations ultimately leading to failure of glands for shafts and bearings). Anti-vibration pads shall also be considered to reduce vibrations.
- Overload settings of electrical starters shall be tamper-proof so that components of the driven machinery will not get damaged.
- Provide guards on susceptible parts such as level indicator tubes; view glasses; and small cocks on steam, water, and oil lines since these can get hit inadvertently and get damaged as a result.
- All instrumentation, safety valves, and rupture discs provided on the equipments shall be calibrated and tested to prevent operation of the equipments beyond safe limits.

13.3.5 Standardise

- Try to standardise the components, items, and spares.
- Check whether same bought-out items are given different names, e.g. bypass valve/flow throttle valve/water line control valve—though all are identical, but used in different sections of the plant. This is to reduce the inventory of such items which are common to many sections of the plant.
- Is it possible to use the same item/spare part for different machines/process units?

For example, can the coupling halves, foundation bolts, drive motors, and covers for different pumps with similar sizes/ratings/design be used for different machines in the plant?
Or, can the level indicator tubes, view glasses, thermowells, gaskets, flanges, etc., with similar sizes/ratings/designs be used for different equipment such as reactors, receivers, mixers, and dissolvers in the plant?

- Joint meetings shall be arranged among shop floor engineers and experienced workers to highlight their problems and give suggestions. These shall be discussed with design engineers from consulting firms and representatives of equipment vendors to modify the equipments so that as many spare parts can be standardised as possible. It may also help to reduce the variety of as many items as possible. Preventive steps such as condition monitoring, lubrication, and preventive maintenance shall also be discussed and their schedules shall be drawn.

13.4 Stocks for Emergency Use/Repairs

Some items like solid raw materials filled in bags or liquids filled in small drums can be stored near the process reactors (but without obstructing escape routes). These can be used for the next batch if the hoist or the feed pump has a breakdown. It allows production runs to continue, while the machinery is being repaired.

Small quantities of spares can be provided near the points of consumption/use to minimise the time required for repairs and replacement, e.g. hose pipe clamps, gauge glass tubes, gaskets, and nut bolts (fasteners). It may otherwise take anywhere from 15 min to 2 h to bring such spares from central stores.

13.5 Suggestions

Some suggestions are given below to reduce the inventory of various items and spares without affecting plant operations and safety. These may also reduce the working capital.

13.5.1 Minimum Stock Level to be Maintained

Minimum stock level to be maintained = Maximum rate of consumption × time required for procurement + some additional quantities as safety margin.

There could be some different views here if the concept of ***just in time*** *is followed in the organisation. However, safety, pollution control, energy efficiency, and product quality shall not be compromised for reducing the essential items in inventory (even if some more funds get blocked due to the additional items stored).*

13.5.2 Reorder Point

An automatic signal shall get generated in the store control system and shall get transmitted by e-mails, general notification, or any other suitable means of communication to all concerned engineers, stores managers, and finance department for warning that new supplies must be immediately ordered to sustain present rate of production. Any further delay may cause slowing down the production or cause problems mentioned earlier.

13.5.3 Minimise the Time Required for Procurement

Minimise the time required for procurement by—

- Working out proper specifications in advance for the materials and items to be procured.
- Regularly floating tenders (if the quantity required is large or of high value).
- Having a ready list of reliable and experienced vendors who shall able to meet all the specifications given in the tender documents, submit samples for checking and confirming the same, and arrange safe transport as per schedule.
- *If the supplier is reliable (who has own manufacturing facilities), it may be possible to procure supplies at short notice and hence lower stock level of inventory can be maintained. It is thus advisable to obtain supplies from a manufacturer directly rather than a trader. It can assure better quality also.*
- Arrangement for cross-checking of samples from independent test agencies or laboratories in case of discrepancy.
- Quickly finalise the technical and commercial conditions. This should be done after considering the landed cost of the items at premises after taking into account the statutory duties (if items are to be imported from another country); and any other costs involved for them (including unloading at site and storage). Take suitable steps to minimise the time required for loading and transport of such materials in suitable containers.

13.6 General Considerations

- Purchase in bulk if being offered at attractive prices (off-season discounts or material quality being better in particular season like summer), on longer duration credit or some other commercially profitable conditions.
- Large amounts of dangerous (toxic, inflammable) materials or those which have less shelf life shall not be generally stored in large amounts. There are chances of accidents, fire hazards, and requirement for elaborate safety precautions and

higher premiums to be paid to insurance companies for such storages. *Inflammable or dangerous materials shall be transported in statutorily approved containers/vehicles having all safety fittings, vents, earth connections, fire extinguishers, etc., only. The route shall be safe and global satellite tracking system shall be installed.*

- Spares shall be procured from original equipment manufacturer and fitted at the earliest to reduce chances of interruptions in plant activities due to breakdowns.
- However, some simple spare parts may be developed in-house, e.g. anti-vibration pads, level indicators, and shaft sleeves for pumps for cost reduction. Sometimes, old spares may be reconditioned and fitted to start the plant soon or if new spares are not available immediately (e.g. impeller or shaft of a blower). This approach may only temporarily save some cost of inventory as a sudden breakdown may still occur.

13.7 Supply Chain Management

Efficient management of the supply chains is necessary to ensure safe, smooth, and energy-efficient running of the process plants while not allowing any environmental pollution to take place—whether the production activities are going on or not.

Release of toxic or dangerous vapours can occur from storage tanks if the cooling system fails for want of some critical spares. This can happen even when the production units are stopped.

Delivery schedules for various products are generally committed by the sales personnel with an aim to maximise the turnover of the organisation, but this can be possible only when the manufacturing units are working properly and *without interruption*.

Sales department shall make commitments to clients (for supply of products) only after confirmation from operating and maintenance personnel that the products will be definitely available in required quantity and as per specifications.

13.7.1 Coordination

There shall be very close coordination among sales, production, stores, maintenance, purchase, and finance department personnel for efficiently managing the supply chain.

Uninterrupted supply of raw materials, additives, catalysts, etc., is required for production, while spare parts are required for timely maintenance. This in turn depends on the stock of usable raw materials and spares available in stores (to quickly repair the machinery whenever required). Spares are also required during annual plant shutdown or during overhauling period for replacements of worn out or damaged components.

Supply chain management for the company will require proper design, planning, execution, control, and monitoring of supply chain activities (while building a good infrastructure and logistics) and matching the supply of all necessary inputs with demand for finished products.

An analysis of the consumption pattern of various items is necessary to arrive at the minimum stock levels to be maintained at all times. A proper estimation of supply of all required inputs as per expected demand should be made.

This can become possible by study of the **actual** operating conditions of the equipments and comparing them with **design**/recommended conditions (temperature, pressure, flow rates, etc.).

Any significant deviation must be investigated and reasons for operating the plant in such a different manner shall be found out.

13.7.2 Example

Analysis of the consumption of high-grade refractory bricks in an electric furnace indicated that the damage was mainly due to higher electric current (amperage).

The operating methods were modified which allowed passage of lesser current (while maintaining the production rate of the furnace) which ultimately reduced the damage to these refractory bricks.

- This enabled to reduce the dependence on special high-grade refractory bricks. They were now procured in the usual manner on reaching the reorder point rather than procuring them as an emergency item.

Management shall therefore design, plan, and establish a supply chain for the management of the procurement and storage of all inputs (raw materials), additives, catalysts, (*for the material undergoing processing*), and finished goods from point of supply to point of delivery or consumption. This can eliminate/minimise manual time consuming procedures such as obtaining quotation, analysis, and approvals of samples, short listing of vendors, and commercial negotiations because timely supplies of necessary inputs at reasonable costs are ensured.

13.8 Planning and Management of Supply

These shall be related to sourcing, procurement, logistics as well as coordination with suppliers and timely information of the current status of all such activities in an integrated manner so that a continuous track can be kept on the movement.

A smooth functioning supply chain will have many links as below.

13.8.1 The Purchase Department

The purchase department has to carry out activities for procuring inputs such as raw materials, additives, packing, colours, stabilisers and materials from regular suppliers, sourcing supplies for more items, negotiation, order placement, incoming transport, storage, handling, and quality assurance, which include the responsibility to coordinate with suppliers for meeting delivery schedules and supply continuity for selected materials.

Procurement process involves development of reliable suppliers who will supply as per manufacturing programme (product mix) to be carried out in response to market changes. There could be urgent demand for the products for which the necessary inputs will also be required urgently. The requirements shall be rapidly fulfilled by suppliers and logistics providers as they form important components of supply chain networks.

13.8.2 Production Department

Production department can assist by planning the quantities and the production of various items to manufacture as per market demand and integrate into the best supply chain flow for the product. An integrated plant can manufacture alum, single superphosphate, 25 % oleum, 65 % oleum, and sulphuric acid. It will need bauxite, alumina hydrate, rock phosphate, elemental sulphur as per demand in market for which reliable suppliers, proper warehousing, and packing as well as dispatch are essential facilities.

13.8.3 Logistics

Logistics which includes packing and loading, transport by railway, ships, road tankers (or through direct pipelines in case of chemicals to be procured from neighbouring units), unloading, and storing at site.

It may also include managing internal processing of the raw materials for producing finished goods and the movement of finished goods out of the organisation to the consumer—who is the final destination, and the timely delivery of the product as per scheduled programme given by customers to make the product available is a vital part of marketing.

13.8.4 *Make or Buy*

Operation of chemical plants requires many items apart from main raw materials and utilities such as additives, property modifiers (anti-caking agents to make the products free flowing, perfumes—which are added to soaps and detergents, foaming agents to tooth pastes and cleaning chemicals; and stabilisers which can arrest slow deterioration during storage).

The products are to be packed in suitable containers before dispatch to customers. Some typical containers are HDPE carbuoys, special glass bottles, bags of various sizes, and stainless steel tanks with suitable lids.

Sealing compounds and gaskets of special designs are required for preventing leaks from lids during transit to distant places or leaks from flanged joints in pipelines in the plant.

Many organisations procure all such items from outside sources rather than producing themselves. The decision whether to make such items in-house or to buy from external vendors can be based on following considerations.

13.8.4.1 Making In-house

Some typical reasons for this decision could be:

- Total dependence on any **one vendor** for an important additive to the product can interrupt production if the supplies are not reliable (not getting delivery on time, in required quantities, is not as per specifications, or the quality is not consistent).
- The cost of buying is more than about 10 % of the cost of production of the main product, and the profit margin of main product has become low.
- If it takes more time, it needs too much efforts to procure from outside.
- It is difficult or dangerous to pack, load, transport from supplier, and unload at site if such items are being procured from outside.
- It is desired to reduce the inventory of such (may be dangerous) items by making them in-house only when required instead of storing large quantities for future use.
- It is possible that the supplies could get interrupted if the manufacturing facilities of the supplier are not running reliably well due to some problem at his end.
- If it is not difficult to produce in-house, and not much resources will be required.
- These items are required regularly, and some of the existing machineries/reactors can be used to produce them since spare capacity is available in-house.
- Not necessary to divert skilled labour and important resources from main plant units.

13.8.4.2 Buying from Outside

Plant Managements can consider procuring certain inputs from outside parties if—

- They are not very critical/important for the production,
- They can be easily available as per specifications given to vendor,
- They can be regularly available in large amounts as per required schedules and at a price which could be lower than cost of producing in-house or only marginally more,
- They can be easily transported or are not dangerous during procurement, and any spillages are not dangerous,
- These are items which cannot be made in-house (*like spares for pumps, conveyers, special lubricants, graphite electrodes, laboratory reagents and other analytical grade chemicals, thermocouples, electrical spares*) since the know-how and/or facilities for making them are not available,
- **Will need diversion of resources or skilled manpower from other important activities if it is attempted to produce in-house**.

13.8.5 Outsourcing

The planned and established specialised supply chain for the companies can enable them to improve their overall competencies. However, in particular, it should allow them to focus on their core competencies—i.e. concentrate on production with safety and pollution control, while supplies of non-critical raw materials (filter aids, salt for water treatment, some stabiliser chemical), common spare parts, and packaging items are outsourced.

The outsourcing helps them to reduce the efforts to procure these items from various sources. These efforts are outsourced to other firms that can perform the activities better or more cost-effectively. The company can thus primarily concentrate on those activities in the production chain in which it has more competencies such as actual production, pollution and quality control, and safety matters.

Hence, the jobs of procuring non-critical items, common spare parts, and their transport, warehousing, and inventory control may be subcontracted to other reliable agencies (managing the logistics). They must be reliable, trustworthy, and able to deliver at reasonable costs.

The effect increased number of organisations involved in satisfying customer demand, while reducing managerial control of daily logistics operations. More supply parties can also ensure supplies through supply chain when there is more than one supplier for the same item.

Main policies can be decided at head office by senior management, while the monitoring and control of supplier performance and day-to-day liaison with logistics partners are best managed by lower level staff or at site.

13.9 Performance Measurement

Reliable suppliers can have a long-term supply chain arrangement with the company, and this can improve customer relationships. Internal benefits are lower inventory cost, better productivity, and enabling attention to more important tasks like safety and quality.

They can help build a better image, reputation, and relationship with customer.

13.10 Warehousing Management

- Facilities for proper loading and unloading and storage.
  Sufficient precautions shall be taken during transport (and storage at own warehouse) to protect the material from rains, contamination due to dust, and exposure to sun.
  Insurance charges and unloading and reloading expenses will have to be incurred **if this is not done**—and the material can get spoiled.
- Stock management system—Delivering products as per specifications and quantities required by clients and maintaining stocks of raw materials required for production of these items.

13.11 Just-In-Time (JIT)

- It can save cost of inventory and space required and facilities required for safe storage. It needs very reliable suppliers who shall be preferably near by the premises.
  JIT is followed when a large inventory can block a lot of working capital.
  Procurement of various items is planned and executed in such a manner that they arrive at point of application *just* when they are to be used/fed into the system. This needs very accurate estimation of the rates of consumption; time required for procurement (testing samples at vendor's works, loading, transportation, and unloading at site) is necessary for the success of this arrangement.
- However, a reasonable stock of the spares shall be available in the stores as matter of abundant precaution.
- There is very little margin available for any delay at any point in the whole chain of supply. There can be a chance of interruption in plant operation due to some unforeseen circumstances such as transport disturbance or some items not meeting specifications (and needing replacement by correct ones). There could be chances of development of unsafe conditions (if thin gaskets are supplied

instead of thicker gaskets for oleum lines) or environment pollution taking place (lime not supplied in time can cause escape of acidic effluents from the premises).
The items shall not be very critical for the safe and pollution-free operation of the plant—hence a little delay in getting them can be tolerated.

- However, this method of getting supplies just-in-time can reduce the cost of inventory in many ways by minimising the area of storage space (especially covered shed), insurance premiums for the stocks being held, and element of danger involved if the large inventories can be dangerous (items which are toxic/inflammable/explosive) or have less shelf life due to hygroscopic or reactive nature.

13.12 Reverse Supply Chain

It is managing the return of materials and spares which cannot be used in the process plant.

Either the supplier shall agree to take them back as such or replace by fresh supplies. This may be costly if the materials are to be sent back over a long distance.

It is necessary that the original approval shall be accorded only after confirmation of the quality by analysing proper representative samples. The original specifications must be defined carefully and explained properly to the supplier before placing orders.

13.13 Some Suggestions Which May Be Useful

- There shall be accurate forecasts of the demand for the products so that steady production activities can be operated. The requirement of raw materials and spares can then be reasonably estimated.
  - *Any excess quantities produced above the demand forecast can be dispatched to warehouses located at strategic places for catering to nearby areas in future.*
  - However, one shall be sure that no product shall remain unsold.
- Having own transport arrangements (own fleet of vehicles or a reliable contractor) also helps to maintain the supply chain.
- Explore possibilities for procuring the supplies from local vendors instead of vendors who may be at considerable distance and therefore need more time for supply.

- But there shall not be any dilution of the essential specifications (which are critical for proper plant operation and product quality).
- Supplies may be obtained in returnable drums or carboys from regular suppliers to reduce cost of packing and to ensure quality of material.

13.14 Improve the Equipment Designs/Operating Methods

Examples:

- Furnaces operating at high temperatures may be provided with better quality refractory linings inside (having higher Al_2O_3 content than before) or operated at lower temperatures (if possible by lesser firing rates of the fuel/lower electrical power);
- Run the agitators in reactors at lower speeds in the reactors;
- Provide flow control/pressure relief valves on delivery lines of process pumps and also provide tripping arrangement for the drive motor to reduce breakdowns by the control of extreme operating conditions;
- Similar steps can prolong the lives of spares as well as the equipments by operating at less extreme conditions (lower temperatures, pressures, speed of agitators).

These changes shall be carried out only after consultation with plant operating engineers, and these shall not disturb the production activities.

13.15 FIFO—First In First Out

- Items stored for long time can get rusted or obsolete and may become unusable. Such items may need reconditioning or repairs before using them or in rare cases may have to be scrapped. This can be a direct loss and must not be allowed. Hence, the items procured **first** shall also be taken out **first** for use.
- However, in a large store, there could be practical difficulty in tracing out the items procured initially, if the procurement activities are going on continuously and new materials are added almost everyday. Hence, a very systematic arrangement for storage is essential to prevent such matters.

13.16 LIFO—Last In First Out

This may have to be followed if:

- There could be too many items already stored and it may take too long to trace out the items procured initially (due to some rearrangement work going on in the stores).

An item/spare with a modified design has been brought, and it is to be tried out immediately (perhaps it is cheaper or has a different material of construction, or another vendor is to be given a chance).

13.17 LILO—Last In Last Out

- If the items can be stored for long time without getting rusted, obsolete, or unusable due to any reason, those procured last can be taken out later on.

Chapter 14
General Administration

14.1 Communications

It is very important to have clear unambiguous communications for safe and smooth commissioning, operations, and maintenance of the plant. Communication means a common understanding of ideas between the sender and the receiver, and for proper understanding, the message must be clear and receiver must interpret in the way the sender intended.

Management must ensure at all times that there is no confusion or misunderstanding among personnel as it can cause accidents, damage to equipments, environmental pollution, and/or loss of production. It is always worthwhile to send the information in the written form rather than oral. This can avoid further misunderstanding in the future. One should use the fastest way of communication processes such as e-mail and video conferencing while communicating.

Various types of Communications are as follows.

14.1.1 Internal

Between process and maintenance personnel (engineers and workforce) or between technical and commercial staff members.

Oral (during meetings/by telephone) or written (e-mails, messages delivered on mobile telephones, personal letters, or by general circulars).

Announcements over public address systems—this could be during incidences of fire or any need for evacuation. This should be as clear as possible and repeated in a steady and calm tone.

Written

Personal or generalised (notices put up on notice boards, published in house journals, in newspapers, professional journals).

K.R. Golwalkar, *Production Management of Chemical Industries*,
DOI 10.1007/978-3-319-28253-4_14

14.1.2 External

External—(with statutory authorities, consultants, vendors, clients)

- **Notifications** for licensing fees, deposits to be made for obtaining power and water connections, and inspections of boilers, electrical installations, and fuel storages are received from statutory authorities and these must be immediately complied with.
- **Clarifications** are also required by statutory authorities before granting consent to start operations (generally these pertain to safety precautions, pollution control equipments, electrical power required for initial construction, and continuous running afterwards).
- Technical reports by consultants—basic and detailed engineering documents and clarifications sought/call for assistance by plant engineers.
- Offers by suppliers for various items required for running the plant.
- Quotations by suppliers in response to inquiries floated—these mention the technical features, prices, and commercial terms. Purchase teams shall study them carefully and look for any technical feature not included at all (but specified in the inquiry), some hidden costs or terms which may not be acceptable.
- Rated capacity, maximum operating pressures, temperatures, and rotational speed of agitators must be clearly mentioned.
- Supply of spares for smooth operation for two years and recommended list of spares.
- One of the most important communication documents is where the performance guarantee terms and conditions are included.
- Exclusions from scope of supply.
- Manuals for erection and commissioning of the plant are prepared by consultants—except for specialised proprietary items or machines. The latter are supplied with the items. The scope of supply by vendor and obligations of the purchaser shall be very clearly defined (e.g. *keeping the civil foundation, water, and power supply and instrumentation air ready before the machinery is delivered at site*).
- Certified copies of tests carried out and results obtained must be obtained.
- Instruction manuals for starting, running, and stopping the machinery as well as specific precautions have to be taken.
- Procedures for overhaul during annual shutdown.

14.1.3 From Superiors (Seniors) to Subordinates (Juniors)

- Orders for product mix to be produced, for keeping certain items ready.
- Instructions on a day-to-day basis (for operating the plant in a particular manner).

- Call for explanation—if something has gone wrong, and a chance is given to explain the situation.
- Standing instructions—to be always followed.
- Modifications in the operating manuals supplied by vendors.
- Standard operating procedures.
- Dos and don'ts—a set of instructions to be scrupulously followed for safe, efficient working. These generally shall include monitoring process conditions, safety precautions, keeping watch on high-pressure process reactors, levels in overhead tanks, etc.
- Refresher courses arranged by the organisation revise the knowledge and improve employee morale and create more awareness about responsibility.
- Supervisors are also able to give clear guidelines and directions for quality and necessary information to make the employees clear about total quality management.
- Requests for feedback in the form of comments on new procedures and modifications proposed in old machinery which can be taken up during HAZOP meetings by experts since a large number of persons cannot be invited.

14.1.4 From Subordinates (Juniors) to Superiors (Seniors)

Lower level of employees should be able to provide suggestions to seniors/higher management. The supervisors shall listen effectively to correct the operations which can then improve the quality of product and working in the organisation.

Suggestions for improvements should be brief, give benefit–cost analysis or convenience in plant operations possible, as well as the time required to implement. Any criticism of existing system shall be impersonal and constructive only. It can create an atmosphere of trust between supervisors and employees.

- Reports on working of plant—production figures, consumptions of raw materials, and utilities.
- Explanations for plant stoppage, breakdowns, or not meeting expected production rates/product quality, and *requisitions made for urgent supplies.*
- Articles published in in-house journals giving some suggestions/experience on job or for building up better relationship among employees. These can improve the morale and prestige which ultimately adds to output and results in more efficiency.
- Presentations for meetings with senior officers to draw attention towards important issues concerning safety, modifications/innovations.

14.1.5 *InterDepartment (Sideways Communication)*

This type of communication takes place in the organisation among the personnel of the different departments:

- Information asked by the purchase department to the maintenance department regarding specifications of spare parts for which requisition has been raised or some comments on the offers received from vendors;
- Electrical department wants to schedule the maintenance of major electrical equipments so they inform their planning to manufacturing or production department.

Intra department: this type of communication takes place within the personnel of same department.

14.2 General

Colour codes for pipelines—It is advisable to follow the codes as specified in American National Standard Institute or Indian Standard 2379 (1990) or their equivalent standards in the country in which the process plant is located. The base (ground) colour and the colour bands shall be clearly visible from different angles to the plant operator and even during night-time. The lighting shall be arranged at suitable angles (light blue for water and orange for inflammable or chlorine).

Signs—skull face and crossed bones—for indicating danger.

Red light—drive motor is ON
Green light—drive motor is OFF

Sometimes, these are used to indicate just the opposite in certain locations. Hence, one should check by ammeter, position of handle on starter, and some noise from the motor and confirm the actual condition.

14.3 Advantages of Good Communication Practice

- Correct (as per instructions) product mix is produced instead of some wrong product;
- Raw material feed is controlled and consumption can be well monitored;
- Operating conditions are maintained for safe and efficient operations;
- Preventive maintenance activities can be planned and executed when plant operators take suitable steps to empty out process reactors and clean them prior to maintenance work (especially from inside).

Redundancy—It is useful to avoid mistakes. *There were four reactors in the plant, and reactor no. 3 was to be stopped for maintenance work. The operator stopped the third reactor while counting from the exit end of the plant. Actually, it was the second one when counting from control room side, and it was not to be stopped.*
A clear instruction should have been as follows:
Stop the third reactor as counted from control room side which has green-coloured handrail (while all others had a black-coloured handrail) ***or*** *which has a 15 HP motor for the agitator, while all others have 10 HP motor. This extra information is not necessary (it is redundant) but will prevent mistake.*

- Employee morale improves when suggestions are accepted and they are praised in house journals or publicly by senior management.
- Mistakes can be avoided in future when explanations are discussed with subordinates without insulting them publicly since they shall work more carefully.

14.3.1 To Take Care

- Seniors must repeat important information, instructions, and any such matters which can affect safety, pollution control, etc., and make sure everything communicated has been clearly understood; and there is no confusion.
- Announcements over public address systems—this could be during incidences of fire or any need for evacuation. This should be as clear as possible and repeated in a steady and calm tone.
- Seniors must know when to speak, what to speak, and how to communicate/speak.
- Important figures shall be repeated by writing in words also.
- Signboards must be in big letters and easily visible even in night. Sufficient lighting should be arranged. The signboards should be placed in prominent places which are visited by as many persons as possible.

14.3.2 To Avoid

- Using only short forms or initials instead of names of items—should not be allowed to cause confusion (especially when dangerous chemicals are to be handled);
- Sign or code words used in a group—the meanings can differ in various groups;
- Threatening, insulting, or slang language—can demoralise the workforce;
- Shouting at juniors—can create negative mindset and loose interest in work;

- Scolding a senior person in the presence of juniors—can antagonise senior persons; they will not be respected by juniors. This can make it difficult for them to instruct juniors and get work done from them;
- Juniors and subordinates must not try to show off that they are more intelligent and efficient than their seniors;
- The chain of command shall never be bypassed except in case of extreme urgency. The senior managers shall not issue instructions directly to the junior plant operating personnel while bypassing the engineers on the shop floor;
- Likewise, the junior personnel must first report to the shop floor engineers instead of informing the senior management directly.

14.4 Example of a Standard Operating Practice

Sulphuric acid plant: starting the plant after a short stoppage.

Follow the following starting sequence for resuming operations:

1. Start steam supply in sulphur feed lines, sulphur pump, and spray gun.
2. Start alkali circulation in scrubbers.
3. Start acid circulation in all towers.
4. Adjust dampers in airlines and start air blower on low load—check whether gases are observed through bypass line on sulphur pump discharge to confirm the sulphur feed line is clear.
5. Start sulphur pump on low load—watch spray of burning sulphur in the furnace through the view port.
6. Operate valves in gas lines to raise the converter catalyst bed temperatures as quickly as possible.
7. Slowly raise airflow and sulphur feed to reach required production rate.
8. Boiler feed water pumps to makeup water level in boilers.
9. Cooling water flow for circulating acid coolers to be adjusted to get optimum temperatures at inlets of acid towers.
10. Dilution water addition to acid system.
11. Monitor acid level in circulation system.
12. Start supply of steam to the consuming points when the pressure builds up in the boiler.
13. Confirm water-level controller for boiler is working correctly.

Stopping for Maintenance

Stop the units in following order:

1. Dilution water addition to acid system;
2. Sulphur feeding to furnace;
3. Air blower after confirming there is no unburnt sulphur in the furnace;

4. Water flow for acid coolers (may not be stopped fully if the PHE manufacturer does not permit);
5. Alkali circulation in scrubbers;
6. Close dampers in airlines.

14.5 Fire Fighting

The plant managements must provide a very effective fire fighting system for the plant *for the safety of personnel and plant.* This shall be provided as per instructions from the statutory authorities. Following steps shall be considered for safety:

- Minimise the inventory of all dangerous, inflammable chemicals and fuels in the plant working areas (day tanks in process plants). Main storage tanks, banks of gas storages, and cylinders shall be located away from main plant, furnaces, etc.
- Provide system for automatic cooling water spray on fuel tanks whenever ambient temperature goes high. Dedicated water tanks shall be available for this purpose.
- Make proper arrangements to store separately all non-compatible chemicals. Their names shall be prominently displayed on signboards and on walls of warehouse.
- Inflammable materials shall be stored at isolated spots. No source of high electrical voltage (transformer; air circuit-breaker, oil circuit-breaker, main panels) or high temperature surfaces (furnaces, steam lines, hot gas ducts) shall be nearby.
- Do not work with naked flames near fuel tanks and gas storages. Install flame arresters to prevent travel back of flames into storage tanks.
- Crushing and grinding operations which can produce inflammable dust shall be carried out at isolated spots.
- Information about flashpoint, vapour pressure, and boiling point for all dangerous chemicals being handled and stored shall be given to persons working and shall be prominently displayed.
- Comply with requirements of insurance companies regarding fire fighting facilities to be provided on the premises, e.g. cast iron pipelines, dedicated water pumps of special design for fire fighting, separate power supply feeders/DG set, CO_2 cylinders, sand buckets, dry chemical powder-type fire extinguishers, long hose reels smoke detectors, and sprinklers must be installed and maintained.
- Install automatically operated audio alarms.
- Dedicated water tanks, (ground level and overhead tanks), standby power supply through DG set, and public address system must be available.
- All these arrangements shall be inspected and approved by the statutory authorities and insurance companies.

- Implement instructions from factory inspectorate always.
- The fire fighting system must be as per approved disaster management plan.
- Conduct mock drills regularly to train all personnel at site.
- Ambulance, first-aid facilities, and oxygen breathing unit shall be always available.
- Neighbourhood evacuation plan and vehicles shall be in place.

14.6 Logistics

The general administration should look after this very important support activity for transport of incoming materials and spares and dispatch of products. There shall be close coordination with purchase, stores, production, maintenance, and marketing departments.

A list of all inputs to be procured and all dispatches to be made in the next few days/week shall be prepared in consultation with the purchase, stores, production, maintenance, and marketing (sales) departments in advance. This should have following details:

1. Names and code numbers of all items.
2. Quantities of each item.
3. Type of packing for each—loose powder or lumps, drums, carboys, bags, glass bottles, or dispatch by tankers/trucks/railway wagons. Special sea worthy packing will be required for dispatch by ships. Very stringent rules are to be followed for sending any item by air (consult export agencies).
4. Which items are inflammable/toxic/dangerous and their quantities in each consignment.
5. Special vehicles or containers required, if any. These shall have statutory approvals.
6. Instructions for safe loading, transport, and unloading of all inputs and outputs.
7. Information about any route not to be followed—passing through inhabited areas, difficult mountain passes, and rough road.
8. Documents which should be provided with each consignment—material safety data sheets, safety precautions, instructions on how to attend spillages, leaks, and antidotes to the items being handled—especially if they are dangerous chemicals.
9. If the clients are located at different locations, it may be convenient to establish depots at suitable places to supply the smaller quantities required by individual clients. This can enable bulk transport from the manufacturing site to the depots instead of sending materials to each and every customer. The landed price to customer can also be reduced.

14.7 Compliance with Statutory Matters

There are various rules and regulations in every country for establishing and operating industries. They are aimed to ensure safety of personnel (inside as well as outside manufacturing premises) and plant machinery, minimising environmental pollution, etc.

These are very stringent in most of the countries.

Some typical statutory matters are given below, but this list is only indicative.

Managements shall make themselves familiar with all such matters along with any amendments coming into effect in their own country and comply with instructions given by statutory authorities from time to time.

14.7.1 Technical

Regular testing and maintenance of all[1] pressure vessels and pipelines, fuel storages and gas pipes, electrical overhead travelling (EOT) cranes, hoists and chain pulley blocks, fire fighting equipments, provision of gas detectors for inflammable, and toxic gases are to be carried out as per instructions issued by statutory authorities and inspection personnel deputed by insurance companies.

14.7.1.1 Plant Layout and Structural Stability

Process units and machineries shall be installed in such a manner that persons present nearby will be able to move away quickly in case of any fire, gas leak, and explosion.

The installations on each floor of the process plant shall have **more than one** escape staircase if the nearest one to the accident spot cannot be used due to smoke or gas leak. It should be possible to rescue the personnel entrapped in the affected area through the alternative escape route which shall have good lighting and ventilation.

<u>Storage tanks and sheds for fuels and dangerous materials</u>

These shall be generally located at isolated spots in the plant so that any fire or mishap will affect minimum personnel and plant equipments.

Fuel tanks and gas lines must have provision of fire fighting systems, CO_2 cylinders, dry chemical powder cylinders, and water sprinklers for cooling them.

The area shall be cordoned off with no entry to vehicles having diesel/petrol engines (as these can emit sparks from exhaust pipes).

[1]It will include all those process vessels also which are likely to get pressurised due to any reason (heating by steam, internal reaction, back flow from connected pressure vessels etc.).

The tanks must have dyke wall enclosures to arrest any leak, and arrangement shall be available to pump out the leaked out liquid. An empty tank is to be always kept available.

Provision of vent scrubbers and gas detectors shall be made. The surrounding concrete floors shall be smooth and sloping with storm water drains. *This will be checked by the technical representatives of insurance companies also to determine the premium amounts and the compensation in case of any mishap.*

Flare towers for burning inflammable gases safely and releasing the flue gases at a height shall be installed.

Carry out regular inspection of structural integrity of all load bearing and support columns/ beams (if heavy machinery is installed on upper floors). There is possibility of vibratory loads due to process units such as agitated reactors/crushers, compressors, grinders, big conveyors, and reciprocating pumps getting transmitted to such structural members. All steel bars and members of support structures shall have corrosion-resistant paint.

The weights due to overhead process tanks, water tanks, or cooling towers of large water-holding basins shall also be considered and *as if they are full of their contents.*

All such inspections shall be regularly done in-house or by appointing licensed engineers and offered for inspection by statutory authorities (factory inspector, electrical inspectors, boiler inspector, etc.) on the scheduled dates.

14.7.1.2 Electrical Items Generally Covered Under Laws and Subject to Inspection

- Main incoming high tension lines, circuit-breakers, and metering panels,
- Installations of transformers and fire fighting arrangements at critical spots,
- Maximum demand and total installed load,
- Flameproof motors, starters, and lighting fixtures,
- Continuity tests for earth connections provided on process vessels (carrying explosive, inflammable, or toxic materials).

The above list is only indicative. Plant managers must comply with laws in their own country.

14.7.1.3 Pressure Vessels

All equipments such as steam-generating boilers, economisers, process vessels operating at or likely to develop high pressure during operation (or even when idle—such as fuel tanks or storage vessels having volatile solvents), compressed air receivers, storage vessels for gases, high-pressure pipes, and gas and fuel carrying pipes are covered by the laws and regulations for pressure vessels in all countries.

They are to be offered for inspection by statutory authorities (*in a cleaned condition after removing all insulation*) as per instructions on prescribed dates. Sufficient safety devices are to be provided by law on each equipment which operates under pressure or is likely to develop pressure due to any reason, which stores or processes dangerous materials which can spill out or get ignited due to any reason.

Such dates must be kept in mind while planning production programmes. *The dates may be postponed in exceptional circumstances (if an urgent order for the product is to be met, and the pressure vessels along with safety devices installed on them are tested just recently) with specific written permission from statutory authorities.*

14.7.1.4 Boilers, Economiser, and Steam SuperHeaters

All such equipments generating and handling steam, pipelines for boiler feed water under pressure, and steam pipelines are also covered under the Indian Boiler Regulations. Plant managers shall comply with relevant regulations in their countries.

Inspection—The boilers and economisers are to be thoroughly cleaned both from tube side as well as shell sides by using suitable cleaning brushes and chemical descalants (chemical solutions with inhibitors such as 3–4 % HCl and rodine inhibitor; sulphamic acid). The thermal insulation and cladding *may have to be removed to expose the bare surface* if the boiler inspecting authority asks for a visual inspection.

Services of local agencies specialised in such jobs may be obtained. The fresh solution and spent solution shall be analysed during the chemical treatment to confirm that excess acid is not used and steel tubes are not attacked while the scales are removed.

Boiler tubes, which have become thin, warped, or are sagging, are generally advised to be changed. Other pressure parts may have to be changed/repaired as per instructions.

Safety valves are set, checked, and locked by boiler inspector during the inspection and hydraulic tests. These settings shall not be interfered with. Old units/those which have been repaired more than once may be derated—i.e. allowed to operate only at a reduced pressure than before. This is for ensuring safety of all concerned and should be complied with.

14.7.1.5 Pollution Control

Methods of disposal being adapted, maximum limits of quantities of effluents which can be discharged after treatments, the concentrations of pollutants in the treated effluents, and disposal pathways for those effluents which cannot be treated by the industries are specified by statutory authorities in every country.

Material balance calculations for the plant shall be updated and shall be available on demand by State Pollution Control Authorities.

Location of disposal facilities like landfill or incineration of wastes are generally not allowed near water bodies such as lakes, rivers, wells, or underground sources for water.

In case of a landfill site, thorough waterproofing of the bottom is required to be done by special high-strength plastic sheets. Leachate collection wells are to be dug around at four directions at 90° intervals (North, South, East, West) to monitor at regular intervals if any pollutants are seeping through.

The exit gases from the process plants are to be properly treated by using equipments such as bag filters, venturi, and packed tower scrubbers per pollution control laws. Tall chimneys shall be provided to release these treated gases and shall be fitted with necessary accessories such as sampling points, aviation warning lamps, and lightening arresters. The heights of these chimneys shall be as per directives from State Pollution Control Authorities.

14.7.2 Inspection of Transport Vehicles

The rules framed by the Transport Control Authorities shall be complied with for safe transport of materials being procured from suppliers and dispatch of products to clients specially while handling dangerous goods. The rules cover standard fittings and mountings (compressor, fire fighting, pressure and temperature gauges, and earth connection hanging chain) on vehicles, capacity of tankers, and driver's licences.

Instructions for not passing by a particular road (near schools, hospitals, market places) during certain hours of the day must be followed.

14.7.3 Legal and Commercial Matters

The management must pay in time the bill for power consumed by the industrial unit, water charges to the external grids for the supply, as well as licensing fees and any other statutory dues to government/local authorities as per applicable laws in the country. This is to ensure that plant operations can be continued without any problem.

Industries are required to make payment of taxes and levies to federal and local governments and make repayment of borrowed funds as per terms agreed with the lenders. The stakeholders shall also be given reasonable return as per laws in force.

14.7.4 Operating Licences and Labour Laws

Rules are framed for proper compensation to the workforce during active working period, treatment, and payment for any injury while on duty and their retirement benefits. Special laws are enacted for service conditions and leave of absence when women are employed.

All regulations regarding working hours, employee's payment, welfare schemes, provident funds and retirement benefits, health check-ups and benefits, prevention of child labour, and special rules for women employees are to be complied as per applicable laws in the country.

Chapter 15
Process Control, Product Quality, and Research and Development

15.1 Laboratory Facilities

Process control laboratory is a must for safe and pollution-free operation of plants while ensuring product quality. It also enables to keep track of the plant operation at different stages of manufacture. The cost of production can be minimised since it becomes possible to trace out the stage of manufacture which is faulty. Managements must invest in a well-equipped laboratory for ensuring the following:

- Analysis of raw materials, additives, stabilisers, and effluent treatment chemicals before their acceptance. The quality of incoming material can be checked at the entry point of the process itself.
- Confirmation that finished product meets specifications given by customer.
- Confirmation that the pretreatment (drying, calcination, filtration, etc.) of raw materials has been done properly before feeding to process plant and that the process is in control at every stage by checking samples of critical streams.
- Checking the performance of process units during commissioning, plant stabilisation, and performance guarantee run. *These are very important stages in the setting up of the industry as they are specifically mentioned in the agreements with the vendors/consultants.*
- Safe working (when processing dangerous/toxic chemicals) by confirming the compositions of the reaction mass for every batch.
- Assistance for troubleshooting problems by examining the process streams of plant or the exit streams from plant when unexplained consumptions of materials occur.
- Investigating process streams for abnormal corrosion of plant equipments.
- Effluent analysis (before and after treatment) for pollution control.
- Development of new process/new products.
- Development of treatment schemes for any product which is not meeting specifications and is to be recovered or disposed off properly.
- Determining reasons for improper product quality.

K.R. Golwalkar, *Production Management of Chemical Industries*,
DOI 10.1007/978-3-319-28253-4_15

15.1.1 Investment in Laboratory Apparatus/Equipments

Investment in laboratory apparatus/equipments which give accurate results in the shortest time is a must. The laboratory apparatus/facilities required shall be procured in consultation with the shop floor personnel.

Many industries have *online instruments for monitoring process streams* (for automatic process control). Readings shown by these instruments shall be cross-checked by laboratory analysis of samples drawn at regular intervals to operate the plant satisfactorily.

Apparatus for qualitative and quantitative analysis. *More apparatus can be added if required:*

- High-volume samplers for stack analysis.
- Electronic balances, magnetic stirrers, and sieves for screen analysis.
- Soxhlet apparatus, condensers, distillation stills, vacuum filters, and bomb calorimeters.
- It is also suggested to provide the power supply to sensitive instruments through voltage stabilisers only, with a backup battery.

15.2 Product Quality

The quality of product should be suitable for actual application. Very stringent specifications given by clients may not be necessary and will only increase the cost of the product. Hence, manufacturer's representative shall visit the client's works and understand how the product will be used. Necessary advice may be given to the client on correct method of use. Discussions with clients can lead to at an acceptable specification (analysis) of the product.

15.2.1 Visits to Clients' Work

Rejection of products by clients can be reduced by advising them on the following:

- The proper way to store (away from dust, sun, rain),
- The correct way to use (add small amounts to reactor at a time instead of large quantities by suddenly emptying of number of bags),
- Take only DM water/good quality solvents instead of hard water/cheap poor-quality solvents for making solutions of the items supplied, and
- The proper sequence of addition of *our* product to *their* process units.

Typical specifications are maximum moisture content allowed, minimum purity required, maximum permissible contents of insoluble matter and other impurities, range of boiling point, specific gravity, appearance of crystals, colour, free-flowing property required, etc.

Certain specifications are however a must—e.g., iron content shall not exceed 15 ppm in battery grade sulphuric acid; insoluble matter shall not exceed 0.2 % in papermaker's alum.

Sampling procedures shall be acceptable to both the supplier and the client. The laboratory analysis results shall also be acceptable. In case of doubt, a sample shall be analysed by a laboratory acceptable to both. Generally, the samples shall be drawn in the presence of representative of client and preserved for future reference.

15.2.2 Factors for Ensuring Proper Quality

15.2.2.1 Process and Equipment Designs

Process and equipment designs should be such that there is minimum possibility of side reactions/formation of by-products (since these can spoil the quality); it shall be easy to separate out unreacted materials or the by-products.

Use of oleums for sulphonation can produce spent sulphuric acid as by-product, which needs to be separated and disposed off.

There shall be no idle pockets/crevices in the equipment where material from previous batch can remain stagnant and contaminate next batch.

Provide air filters at inlets of blowers to remove moisture and dust from air streams to be used for aeration process or for the reaction section *where intermediates are being manufactured for supply to pharmaceutical industries.*

These filters minimise corrosion and erosion of impellers or other internals. The dust is also prevented from entering downstream units as it can spoil the products.

Provide dust control equipments for crushing and grinding units since the dust can enter electrical circuits, junction boxes of motors, any openings in starters or relays, and cause short circuits leading to breakdowns. The frequent plant stoppages can disturb process settings and the product quality.

The dust can also enter reactors through charging manholes or any small openings and contaminate the product.

Traces of leftover process materials such as catalysts, anti-caking, and anti-foaming agents from previous batch can spoil next batch. Hence, process units such as reactors, heat exchangers, and filters shall be easy to drain out and clean for next batch.

Charging manholes, agitator shafts shall be provided with gland packing of compatible materials or shall be water-cooled type to prevent escape of vapours. Mechanical seals may be fitted in case there is a possibility of escape of toxic or inflammable vapours.

15.2.2.2 Materials of Construction

Materials of construction of reactors and agitators, protective lining of reactor, and catalyst bed support grids shall not react with the reactants being processed.

Use stainless steel as material of construction for process vessels, pipelines, storage tanks, heat exchangers, etc., in order to minimise contamination by iron compounds.

Absorption/distillation towers shall have packings with glazing/non-corrosive material to prevent contamination of process streams which shall be tested for compatibility with process fluids at operating conditions (e.g. the tower packing in any of the drying, interpass absorption or final absorption towers of the sulphuric acid plant shall not get affected by 98.5 % sulphuric acid at 120 °C).

The refractory and acid-resistant lining of vessels shall be inert towards the materials being handled. Surfaces in contact with the reactants shall be corrosion- and erosion-resistant. They can be applied with a layer of screened cpharmaceutical industries or if it forms part ement paste and made smooth.

The gaskets at joints in pipelines can be made of Teflon, neoprene after testing in the laboratory at the operating conditions to confirm they will not react with the fluids.

The material of construction for reactors, process vessels, heat exchanger tubes, and process pipelines can be similarly chosen from carbon steel, suitable grades of stainless steels, Teflon-lined steel, fibreglass, polypropylene, glass, etc., after laboratory tests at operating conditions to confirm suitability (for minimising chances of contamination).

15.2.2.3 Design of Plant Layout

It should be convenient and easy to observe the levels in tanks, pressure and temperature gauges, and length and colour of flame in furnaces to keep a watch on the running of plant units.

Ease and safety during operation and maintenance, proper supports and staircases, work platforms, shadowless lighting, railings around the reactors, and process vessels to prevent accidental fall inside the vessel—all such steps improve the confidence and motivate the working persons for paying more attention to the operations.

Storage of raw materials with isolating bays for similar-looking items, proper labelling, and sign boards with display of stock levels can reduce mistakes in operation (e.g. feeding of wrong material to process).

Storages shall be safe with provision of dyke walls to contain any overflow or spillages with arrangement to recover the material and arrangement to neutralise if necessary (when recovery is not possible or the recovered material may contaminate the material in the main storage tank if again put back).

Ensure proper lighting, escape routes, safety vents with release points away from working areas, piping routes which will not hinder movement of personnel or accidently hit body parts, good external insulations, exhaust hoods, and induced draught fans to remove fumes from working areas. All these steps add to the efficiency of the workforce and result in better product quality since the process units can be paid more attention by them. The process conditions can be monitored and controlled in a better manner.

15.2.2.4 Raw Material Quality

Procure only those raw materials which meet the specifications given by production engineers.

Manufacture of sulphuric acid

The presence of some harmful contaminants (e.g. ash, organics) can reduce activity of catalysts in sulphuric acid plants, or corrode equipments and affect product quality.

Manufacture of white alum for paper

- *Presence of iron compounds in ground bauxite imparts yellowish colour to alum (hence use alumina hydrate for pure white colour).*
- *Do not use spent sulphuric acid; use only fresh acid with iron content below 15 ppm.*

Any improper functioning of purification processes (filtration, screening to remove tramp materials, calcinations to remove volatiles) can affect product quality.

Use *demineralised water instead of filtered water* for making solutions, where water is to be used in pharmaceutical industries or if it forms part of final product, **e.g. copper sulphate crystals $CuSO_4$, $5H_2O$; manganese sulphate crystals $MnSO_4$, H_2O**.

15.2.2.5 Additives

Stabilisers—to be added to product to prevent degradation by storing in sunlight, high ambient temperature, ingress of moisture (through vent valve of storage tank which has remained open for a short time). The quantity to be added shall be very small generally and to be used only after testing in laboratory for suitability.

Property modifiers—Anti-foaming and anti-caking (fine silica powder to ensure free flow of product) shall be added after approval by customer.

Retarders—for increasing storage life: such additions shall be done after getting permission of clients and shall not affect performance at user's end. These shall be generally required in very small dosage (less than 0.1 %), and their cost shall be low. They shall be thoroughly miscible with the products and shall remain mixed—*rather than separating and settling at bottom of tanks after some time.*

15.3 Process Control

15.3.1 Some General Checks

- Confirm accuracy of all flow measurement, level indicators, thermocouples, pressure gauges, weighing devices, etc.
- Confirm working of all non-return valves in process lines especially for adding materials to process reactors (backflow of materials can contaminate the materials in supply tanks).
- Calibration of metering pumps to be tested by collecting the discharge in separate tanks. This is to ensure correct feeding of raw materials to process reactors.
- The instruments can indicate correct readings when the probes are installed in flowing fluids and not at idle pockets of vessels or ducts. The probes should be corrosion-resistant or installed in suitable enclosures (stainless steel/ceramic/ Teflon—coated thermowell instead of stainless steel/ceramic/Teflon coated thermowell). Strainers shall be provided in sampling lines to prevent deposits of salts and ash on the probes especially those for pH measurements and concentration analysers.
- Check samples from process streams/at inlet and exit of every unit to properly monitor and control the process at every stage.
- Accurate control of the process shall be done by measurement of flow, pH, concentration, temperatures, and pressures by using instruments and automatic controllers.
- Provide online instruments for continuous indication and quicker control action. *All these must be well maintained and calibrated at regular intervals. The readings indicated by instruments shall be cross-checked by laboratory analysis also.*
- Regularly check that the settings of control valves are correct and there is no gradual drift due to vibrations, high flow, and any moisture or ash deposit.
- Results of five consecutive samples shall not be above (or below) set values—this may be due to drift in the control valve setting.
- Diesel generator sets of adequate capacity to run the plant shall be provided with auto start mode if power failure occurs—this will minimise resetting of process controls and valves. *It will need a large capacity DG set if the entire plant is to be kept running by the DG set when the grid power fails (and not only emergency loads). This arrangement is generally required by industries manufacturing special high-value chemicals or which can get spoilt during even a short interruption of power.*
- Plant engineers can add more such check points for their plants.
- Plant authorities shall draw samples from the product lines. Upper and lower control limits shall be set as per acceptable quality standards and confidence levels.
- **Example**—Dilution water is added through an automatic control valve to the circulation tank of a sulphuric acid plant to maintain the strength between 98.2 and 98.6 % with a mean value of 98.4 %. If six consecutive acid samples show

results more than 98.4 % (but still below 98.6 %), it means the acid strength controller is drifting out of control and the setting for dilution water addition needs to be looked into.

- **Select suitable sampling procedure** when heterogeneous materials are to be examined. Typically, random samples are drawn from the product lots, mixed together, and a composite sample is prepared for analysis. Sufficient number of samples shall be drawn to get proper representation of the product.

Liquids generally have uniform composition due to mixing *(which can be ensured by slow-speed stirrers in the storage tank if required). Higher speed of stirring can disturb the sludge settled at the bottom—hence, care is to be taken.*

15.3.2 Supervision of Plant Operations

Regular monitoring of condition of equipments so that plant stoppages due to break downs are minimised and production runs of the plant are not interrupted—since frequent stopping and restarting the plant can disturb the operational settings of valves and product quality may suffer as a result.

Plant shall be run at as near the optimum capacity where the performance is best possible rather than overloading it (which can increase amount of impurities in final product).

Running it much below capacity can make it very difficult to maintain optimum process temperatures/flow rates in process units such as converters, heat exchangers, and evaporators, and adversely affect running of these equipments and hence the product quality.

There can be heat losses from the ducts and shells of process units, and the temperatures can get disturbed when the plant is run at too low rate.

At very low temperatures, the reaction rates may get disturbed, and at very low gas velocity, the heat transfer will be affected.

15.3.3 Selection of Sample Size for Examination of Produced Items

It is necessary that the items being sold must be as per the specifications given by the customers for getting satisfactory performance at the points of use. It is not possible to check each and every piece of the product if it is produced as a separate individual piece on a large scale. A certain criteria must be evolved for the inspection (laboratory analysis) in order to have a reasonable level of confidence about the quality. Following are some of the factors which can be considered for developing these criteria for acceptance.

The above reasoning is also applicable to items being procured as raw materials and other inputs for maintaining the quality of the products being manufactured.

- Intended use of the product which is being sold to clients.
- Point of application of the item which is being procured from outside.
- Number of items being sold (or being procured).
- Price of each item.
- Consequences of an item not meeting required specification (e.g. laboratory reagents must be very pure; otherwise, the results of analysis shall not be correct).

  Consequences of the length of writing chalk being more or less by 4–5 mms will not be serious. Similarly, consequences of the quantity of cement being less or more by 50–100 g in a bag of 50 kg are not serious.
  But a difference of even 5 g in a vane of a high-speed impeller can imbalance it and damage a costly mechanical equipment such as a turbine or an induced draught fan. Likewise, an impurity of even a few ppm can spoil a pharmaceutical intermediate.
- **Size of each lot**:
  Purchase Department shall check whether the items are being produced regularly as a standard product in large numbers, e.g. rectangular-shaped refractory bricks of size 230 mm × 115 mm × 100 mm are generally produced regularly in large numbers with alumina content of 35–40 %. These are used for brick lining of flat or straight surfaces.
  However, specially shaped refractory bricks with curved surfaces (used for manhole arches or for lining circular shapes of furnaces) or with high alumina content of 65–70 % are manufactured only as per special order placed by clients in limited numbers.
- In the case of standard rectangular bricks, the dimensions are not very critical and a little deviation can be tolerated. Hence, it shall suffice if only about 40–50 bricks are examined at random, out of a lot of say, 2000 bricks. However, in case of the curved bricks, they must have correct dimensions for proper installation (a tight fit) so as to obtain a good strength for the arches. This calls for a much closer inspection by a higher sample size from the lot for examination, and at least 100 curved bricks shall be examined from a lot of say, 1000 bricks. One sample may be drawn after every 10 bricks for inspection.
- Another example is that of acid-resistant Raschig rings which are required in thousands of numbers in process plant absorption towers while only a very small number of candle demisters (2–8) may be required for acid mist control. Here, only about 10–15 Raschig rings are tested from a lot having a few hundred numbers. But, it is necessary to thoroughly examine *each and every* candle demister before dispatch and installation in the towers. The reason is that a few Raschig rings with a little deviation in dimensions will not affect the performance of the absorption tower, but even a slight deviation in the candle demister can drastically affect the performance of downstream units due to escape of acid mist particles.

15.3.4 Shelf Life of Items

Those items (e.g. gas mask canisters, and personnel safety devices such as face shields and safety goggles) which are likely to deteriorate during long storage need to be checked once again before actual use. Some more examples are electrical spares, catalyst used in reactors, spares of instrumentation, anti-caking agents, etc. If a proper inspection of such items is not done before use, it can result in serious problems in the process plant.

15.3.5 Confidence Level Required

To check whether it is desired to be absolutely sure that not a single item is defective since the consequences of using a defective item can be very costly (Examples—impurities in AR-grade chemicals can result in wrong results or using a defective unbalanced impeller of a steam turbine may even result in accidents).

- or
  Whether it will be sufficient to have a 90 % level of confidence, i.e. one is sure that the product will meet the required performance 90 % of the time *if the application is not very critical (e.g. galvanised steel pipes or rubber hose pipes for water service when it is not for a very critical application)* **or** when alum is to be used for only an initial settling treatment of muddy water.
  The draw of samples of the various items such as special plastic containers or glass bottles shall be such that a sufficient confidence level is necessary before filling them with products such as acid or oils. Any leakage from such container can be dangerous. Hence, all the containers shall be inspected in such case.
- In case of other items which are not to be used for very critical application, a sample can be drawn after every 5, 10, or 15 pieces from the assembly line **or** randomly drawn from the manufactured lot and tested. If the sample is able to meet the specifications, then the lot is generally accepted. The reader is requested to refer to information available on the internet for the sampling plans for various confidence levels.
  An important consideration is whether the product is a solid distinct piece or a liquid filled in storage tanks. The liquids are generally miscible even from different batches and can form a homogeneous lot in the storage tank. Only a few samples from this lot may be analysed to get correct results.
- Generally, the clients give the specifications as a numerical value for acceptable quality (minimum purity %, solubility %, viscosity as centipoises). While a product may meet the criteria for minimum purity, it is possible that it may contain a component which is not acceptable.
  Example insoluble content not to exceed 0.2 % in papermaker's alum even if it has 16 % Al_2O_3 as specified (for acceptance).

or Iron (Fe) content not to exceed 20 ppm in battery grade sulphuric acid even if the acid is of 98.5 % strength as specified (for acceptance).

- Such tolerance limits are the maximum extent of impurity the client *may tolerate. However, this does not mean the client will definitely accept a product with 18 ppm of Fe content. They may ask for more samples to be analysed in such case.*
- Generally, 3 sets of samples are drawn. Each set may consist of about five to ten samples drawn at random from the lot (items to be sold or being procured). The different samples are analysed individually or they are mixed thoroughly to make a composite (average) sample.
- Mean value of the set, individual values, and maximum deviations are noted.
- One set is analysed in the supplier's own laboratory, the second one is analysed in the purchaser's laboratory, and the third one is kept for future reference or for analysis by an independent laboratory in case there is a discrepancy between the results of the supplier's and purchaser's laboratories.

15.3.6 Sampling Errors

It is very important that correct reliable analysis of the process streams is available at the earliest for taking suitable corrective actions to improve the process control.

Ideally, this is possible with online sampling. But, this may need costly instrumentation and can be difficult to maintain. Manual methods of drawing samples and analysing them in laboratory could be time-consuming and inaccurate. Some of the reasons for this could be as follows.

In case of solids, it is important that small amounts of samples are randomly collected from various lots of the stored products prior to dispatch. These shall then be mixed thoroughly to get a representative sample of the product. The results can be unreliable regarding the quality if sufficient samples are not drawn.

In case of liquids, a common mistake is to draw the samples from wrong positions such as idle pockets in a flow channel, corner of a rectangular tank, or through a very long sampling pipeline. The time taken for obtaining a correct representative sample from the flowing liquid is volume of sampling pipeline (litres) divided by rate of draw of sample (millilitres per sec).

A long sampling line (with a small diameter) can cause a time lag in the corrective action because the parameter being checked has changed by the time the results are known and corrective action is initiated.

In practice, one should take care of the following for proper corrective actions:

- Minimise the time required for obtaining representative samples by having a short sampling line from the main pipe through which the fluid is flowing,
- Preconditioning of sample before analysis is done (e.g. removal of acidic mist particles is a must from an exit gas stream of a sulphuric acid plant being checked for SO_2 %),

- Use fast laboratory techniques and suitable apparatus for analysing the sample,
- Provide suitable feedback controller or feed forward controller for carrying out corrective actions in plant process units. The type of controller depends on the process parameters and the system configuration. The matter shall be discussed with instrumentation engineers, shop floor engineers, and vendors of controllers,
- Another source of error is drawing samples *intermittently* from the flowing streams. It is possible that the sample is being taken out when everything is alright while no sample is drawn when the process is not running properly. It can create a false sense of complacency that the process is under control (e.g. *intermittent sampling of treated effluent streams may indicate that the pollution control units are functioning properly because the analysis of the treated sample may be satisfactory even if the actual conditions are otherwise*).

15.4 Material of Construction of Vessels and Containers

- MOC of tankers, bags, and carbuoys should be compatible with the products.
- Products shall be packed/filled in containers made from non-contaminating material of construction such as high-density polyethylene (HDPE); lined with poly-tetra fluoro ethylene (PTFE), stainless steel 316.
- There shall be tamper-proof seals and leak-proof lids.
- Confirm that the operations for filling of containers and sealing properly are being strictly supervised, so there should be no spillage or difficulty during use.
- Accurate measurement of quantities being sold must be ensured. Install metering tanks for sale of liquid products. Any short supply to clients can create bad impression in market, while excess supply can erode profits.
- Check use of tankers—whether they were used for some other chemicals earlier and were not washed properly. *Tankers shall be used for only one product.*
- Hence, the organisation should have own fleet of tankers.
- Product tank internal cleaning shall be done at yearly intervals (or earlier). The tanks can have lining by acid-resistant bricks if corrosive products are to be stored.

15.5 Innovation, Research and Development

Progressive managements shall consider these activities as very necessary for more safety in operation and maintenance, better quality of the products, getting leadership in market, increasing profits, more overall efficiency in manufacturing, sometimes even for survival of the company (if there is very stiff competition for selling the products or if better/cheaper products are available), and for diversification to other products.

- Innovation is for improvement in existing set-up.
- Research is for new process.
- Development is done for new technology (could be based on known process or based on new process).

Innovations in equipments or technology shall always be discussed among senior engineers and experienced workforce initially and the benefit—cost analysis worked out. Even though a large benefit appears likely as a result of the innovations, it may be convenient to introduce only small changes in the plant at a time (so that the plant operations are generally not disturbed). The effect can be observed over a reasonable time before making the next change. This will enable the management to come to a logical conclusion to confirm the change which resulted in more improvement at lower cost.

Some typical innovations could be changing the speed of agitation in a reactor, changing the cooling water supply a little bit (say, 5 %) to the process, changing the sequence of additions of raw materials to a reactor, increasing the amount of catalyst slightly, etc. The effect of these small changes is to be monitored, and if better results are obtained, further changes may be carried out.

It may be calculated *if possible* what are the best performance figures theoretically possible for manufacturing such products?

The following shall be then discussed in-house with the technical and commercial teams:

- What is best possible in given circumstances with present knowhow and equipments available (i.e. with current technology being used) if the plants are run and maintained in the most efficient manner?
- What is the present status? How much is the difference from the best performance?
- What needs to be done to overcome the difference?
- Whether entirely new machinery is required or some modification/upgrading of existing equipments can be sufficient for the purpose?
- How much time and funds will be required for this purpose?
- What are the chances that the present technology may become obsolete? The efforts made for upgradation and the costs incurred may then go to waste. Will it be acceptable?
- Are competitors making better or cheaper products and flooding the market?

In above situations, it is advisable to invest in research for new process and allocating fund for development of technology—*which should result in a safer process for manufacturing of the products with better quality, having a better environmental pollution control and more overall efficiency (lesser raw material and energy consumptions, more equipment life, better use of manpower, etc.).*

The newly developed process should also reduce dependence on the externally procured (imported) materials and catalyst if they are being used presently.

- However, it shall be possible to use as much of the presently installed items and present machinery to minimise additional investment.

- It should also become possible to use cheaper raw materials which can be procured locally—*even though they may not be of the same good quality as before*—(e.g. may have more moisture/more ash content). Some pretreatment may be required in such cases.

Examples for Innovation

A sulphuric acid plant can run efficiently with elemental sulphur as a raw material (having very low ash and organic content), but this is not always available at many locations. The plants located at such places depend on imported sulphur. They can use sulphur with a more ash content (which can be cheaper and available locally in certain countries) with *prefiltration of sulphur and special ring-type catalysts in converter passes which are less prone to getting choked due to ash.*

Use of caesium-promoted catalyst made it possible to operate the last pass of converter at a lower temperature (395–400 °C) as compared to 425–430 °C with conventional catalyst, and this could achieve a higher equilibrium conversion of SO_2 to SO_3.

15.5.1 Aims of the Activities for Research and Development

- Enable diversification to new products using existing machineries and facilities as much as possible or by adding a minimum of additional equipments, so that the company can survive even when there is reduction in demand for the present products.
- Should increase the life of existing equipments by reducing the number of breakdowns and the cost of repairs by operating the modified process/incorporating innovations.
- Result in reducing the side reactions and increase life of catalyst.
- Reduce consumptions of inputs such as raw materials, power, and water.
- Make the operations easier with less manpower and thus allow the rationalisation of manpower which *may become redundant.* They may be deputed to other duties. This can create goodwill for the management as well.

Example for Research and Development

Membrane cell process for manufacture of caustic soda has mostly replaced the mercury cell process due to better efficiency. The mercury cell process also had environmental pollution issues that have been removed in the membrane cell process.

Chapter 16
Smooth Running of the Production Plants

Typical check lists of important equipments which must be carefully operated and maintained in good condition is given below for some plants so that production can be continued smoothly. The production and maintenance personnel shall be instructed to refer to the list every day, carry out preventive maintenance, and attend any breakdowns immediately. The formats are only suggestive for a typical plant for these products. It will be useful if an actual working flow sheet with full details of operations being done in the plant, the actual equipments normally in use, and the standby equipments installed are made by the plant engineers for their own plants. The checklists may be even put up on marking boards on the walls of the control room and will serve as daily reminders to all concerned. These formats may be summarised and discussed in first week of every month in meetings for the production planning by senior production and maintenance engineers, marketing, stores, purchase, and finance departments.

Life Cycle Management

The plant management shall have regular coordination meetings with the sections working for marketing (of products), purchase (raw materials and other items), engineering stores, maintenance (electrical and mechanical), instrumentation, finance, logistics, etc., to ensure proper life cycle of the plant especially when more than one product is manufactured; and there is possibility of better or cheaper products coming into the market. More uses and applications for the products shall be researched to minimise the chances of its becoming obsolete.

16.1 Process for Manufacture of Sulphuric Acid

Sulphur is melted in melters by steam coils. It is then pumped to the refractory-lined furnace to produce SO_2 by burning. Air blower supplies dry air to the furnace through the drying tower. The hot gases from exit of furnace are cooled in a waste heat recovery boiler #1 and then passed through a multistage converter having

K.R. Golwalkar, *Production Management of Chemical Industries*,
DOI 10.1007/978-3-319-28253-4_16

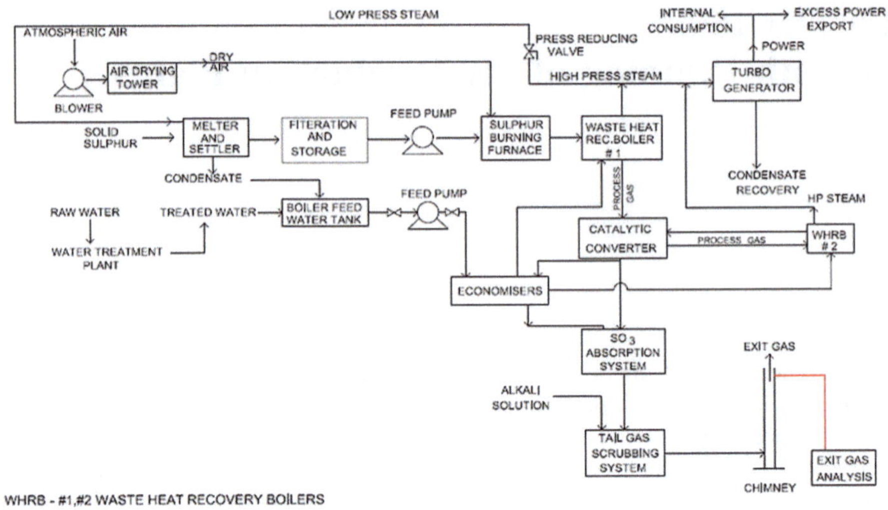

Fig. 16.1 Manufacture of sulphuric acid with heat recovery

vanadium pentoxide catalyst for the conversion of SO_2 to SO_3. Heat evolved during conversion is also recovered by economisers and another waste heat recovery boiler #2. The SO_3 is then absorbed in streams of sulphuric acid in interpass absorption towers (IPATs) and final absorption towers (FATs). Dilution water is added at appropriate rate to the acid to maintain its concentration at optimum value. Exit gases from final absorption tower are passed through scrubbers.

Acid towers, circulation tanks, and coolers are not shown in Fig 16.1 (Table 16.1).

16.2 Batch Process for Manufacture of Papermaker's Alum (Iron Free)

Alumina hydrate (AH) (as powder or in bags) is taken up by a hoist and kept at charging floor. 98.3 % iron-free sulphuric acid is pumped to the day tank. Treated water is stored in feed tank. Measured quantities of acid and water are fed to the reactor at a controlled rate for a particular batch. There is rise of temperature due to the dilution of the acid. AH is slowly fed into the reactor. Steam is supplied (from WHRB or oil-fired boiler) to the reactor coils to complete the reaction. Samples of the product are taken and tested for purity if any unreacted AH is remaining. Steam is kept on for some more time. The hot product is drained out to mould and allowed to solidify (Table 16.2).

Table 16.1 Checklist for manufacture of sulphuric acid

Main equipments	Required for	Consequences of breakdown	Remarks
Steam coils in melter and filtration system	Sulphur melting and filtration	Reduction of production	Keep ready spare steam coils
Sulphur feeding pumps	Sulphur feeding	Production can stop	Install spare sulphur pump
Air blower	Supply of air to furnace (and plant)	Production can stop	Install standby blower
Waste heat recovery boilers	Maintaining gas temperatures at converter passes and for steam generation	Less conversion of SO_2 to SO_3 and less steam generation	Critical units
Economisers	Heat recovery and for maintaining gas temperatures at IPAT and FAT	Less steam generation if not in use due to any breakdown; less absorption of SO_3	Provide another means to cool the gases (air-cooled heat exchanger)
Water treatment plant	For boiler feed and for production of acid	Boilers may come in danger, and acid quality will suffer	Keep treated water stock for 12 h requirement
Boiler feed water pump	Water supply to boilers	Boilers may come in danger	Provide standby pump
Converter system			
Catalyst beds	Conversion of SO_2 to SO_3	Environmental pollution, loss of production	Monitor pressure drops and temperatures every half an hour
Heat exchangers	Conversion of SO_2 to SO_3	Environmental pollution	As above
Acid circulation system			
Circulation tank, pump, acid coolers, water pumps	Maintain proper acid temperature and flow at acid towers	Can stop production and environmental pollution	Provide standby pumps; attend acid leak immediately
Acid towers			
Drying tower	Drying of air	Acid mist generation in downstream units	Check every hour
Interpass absorption tower	Absorption of SO_3	Environmental pollution	Check every hour
Final absorption tower	Absorption of SO_3	Environmental pollution	Provide alarm if exit gases pollute the environment
Instrumentation			Regular calibration and maintenance
Temperature indicators	Smooth efficient operation of the plant	Production disturbed; Environmental pollution	Monitor every half an hour
Acid strength controller		Environmental pollution	Monitor every half an hour

Table 16.2 Checklist for manufacture of papermaker's alum

Main equipments	Required for	Consequence of breakdown	Remark
Hoist	Lifting alumina hydrate (**AH**) bags to charging floor	Production affected severely	Manual lifting of bags may be done
Main reactor with lead lining	Reaction of AH with sulphuric acid	Production affected severely	Immediate repairs required
Boiler with accessories	Supply of steam to reactor and evaporators	Production affected severely	Maintain properly
Fuel tank, fuel oil pump, air blower	To run the boiler	Cannot run boiler	Production will be affected
Water storage and treatment	To run the boiler, to adjust the strength of acid	Production of good quality alum can stop	Do not use untreated water
Steam coils	To complete the reaction and operate evaporators	Production rate and product quality affected	Standby coils to be kept ready
Acid transfer pump and day tank for acid	Filling up acid in day tank for reaction	Production gets delayed if day tank gets emptied	Manual lifting of acid carboys **not advised**
Moulds	To make alum slabs	Production delayed	Keep spares

16.3 Process for Manufacture of Alum from Bauxite

This product is used as a settling agent for water treatment. Bauxite lumps are crushed and ground to a fine powder (80 % passing through 200 mesh). Cyclone and bag filters are used for arresting pollution due to fine dust. Concentrated sulphuric acid (need not be iron free) is pumped to the day tank. Treated water and recycled filtrate (*from filter press*) are stored in separate feed tanks. Measured quantities of acid, water, and bauxite powder are fed to the reactor at a controlled rate. Steam is blown into the reactor to complete the reaction. Reacted mass is diluted by the addition of steam condensate (from evaporators), recycled filtrate, and sent to thickeners for settling of impurities and unreacted material. The clear solution from overflow line is concentrated by evaporation to about 60° by steam coils installed in evaporators. Steam supply is kept on as per need. Concentrated material is checked for quality. It is then drained out to mould and allowed to solidify. Pasty sludge from thickeners is washed with water which is sent to filter press. Filtrate containing some dissolved alum is recycled to process. Washed sludge and filter cake are sent to drying bed for disposal. Steam for the evaporation is obtained from coal-fired boiler or waste heat recovery boiler in the plant (see Fig. 7.1; Table 16.3).

Table 16.3 Checklist for manufacture of alum

Main equipments	Required for	Consequence for breakdown	Remark
Bauxite feeder	Feeding the lumps of Bauxite through a grid of required size	No feeding of raw material to crusher and grinder	Check lump size and moisture % of the bauxite
Crushing and grinding machines	Lumps are reduced to coarse particle and then to fine powder	Oversized particles will cause incomplete reaction	Check particle size by screen analysis
Bag filter for dust control	For the control of air pollution during crushing and grinding	Fines can escape from damaged bags and cause atmospheric pollution	Keep spare bags ready
Acid-resistant AR brick-lined reactor	For reacting bauxite powder with sulphuric acid	Damaged AR lining can corrode reactor shell and it may leak	Check AR lining after each batch
Sulphuric acid, main storage tank, transfer pump, and day tank	Sulphuric acid is pumped to the day tank by the transfer pump	Failure of the transfer pump can empty the day tank which will interrupt production	Carefully operate acid transfer system. Keep spare pump
Boiler with accessories, Fuel storage, oil-firing system, air blower	Supply of steam required for reaction and evaporation of dilute liquor	Reaction mass can get jammed; production and evaporation interrupted	Steam from any other source may be used
Water storage and treatment facilities	Generally treated water is required for both steam generation and mixing purpose	Boiler may come in dangerand may affect the quality of product	Keep a tank filled with treated water always ready for the plant
Thickener (settler and clarifier tanks) for settling of unreacted material as thickened slurry	Clarified dilute alum solution overflows to evaporator	Loss of product alum with bottom slurry if stirring at controlled rate and settling not achieved	Optimise dilution, the speed and design of rake mechanism
Filter press with back wash arrangement and accessories (air compressor)	For separation of un reacted solids/waste and recycle of filtrate	Loss of product alum with unfiltered slurry and possibility of land pollution if filter press does not work	Filter press cake can be disposed off to landfill
Instrumentation Dial thermometers, pressure gauges, pH metre	For monitoring of reaction, operation of evaporators, product quality	Incomplete reactions, poor product quality	Laboratory analysis at every stage to check reaction progress and product quality
Evaporators with steam coils, pressure gauges	For concentrating dilute alum solution by using steam coils	low concentration will result in poor product quality (free acidity may remain); difficulty in moulding of slabs	Keep ready spare steam coils

(continued)

Table 16.3 (continued)

Main equipments	Required for	Consequence for breakdown	Remark
Moulds	For making standard sized slabs of solid alum for sale	Customers may object to broken pieces	Keep spare moulds available
Weighing and packing	For sale of product	Accumulation of unsold product till these are attended	Spare stitching machine for bags
Drying bed	For drying of wet filter press cake prior to final disposal	Land pollution	Dry in shed if sun drying is not possible

16.4 Process for Manufacture of Single Superphosphate

Rock phosphate lumps are crushed and ground to a fine powder (about 90 % passing through 100 mesh) and stored separately. Cyclone and bag filters are used for arresting pollution due to fine dust. 98 % of sulphuric acid from main storage is pumped to the day tank. Treated water is used to dilute it to about 65–70 %.

Both rock phosphate powder and acid are fed separately at controlled rates to the **acid-resistant AR** brick-lined mixer (reactor) with paddles made from special corrosion-resistant alloy. The reaction mass is cured for completion of reaction. Gases evolved from mixer contain fluorine compounds (due to the presence of CaF_2 in rock phosphate). They are scrubbed, and the solution of H_2SiF_6 obtained is recycled to mixer to reduce the consumption of fresh sulphuric acid. It is advisable to select the raw rock phosphate with as little CaF_2 impurities as possible (not more than 3.0–3.5 %) because the very objectionable pollutant HF is evolved during reaction with acid (see Fig. 7.3; Table 16.4).

In addition, it is very important to maintain the material handling equipment such as conveyors, automatic shovels, overhead travelling cranes, bag filling, weighing, and stitching machines in the shed-cum-warehouse where hundreds of tonnes of the product is stored.

16.5 Process for Manufacture of Caustic Soda and Chlorine

Fresh salt is dissolved in recycled (depleted) brine which is filtered and purified by the addition of Na_2CO_3, NaOH, etc., to remove the dissolved calcium and magnesium salts by precipitation. It is fed at controlled rate to the electrolytic cell. Sodium ions migrate through the special membrane to the cathode chamber and

Table 16.4 Checklist for manufacture of single superphosphate

Main equipments and facilities	Required for	Consequence for breakdown	Remark
Storage shed for rock phosphate lumps	Rock phosphate in the form of lump is stored	Wet material is difficult to feed/crush/grind	Do not allow ingress of rain water inside shed
Pulveriser/grinder with air separation and dust control	For pulverising/grinding rock phosphate	Production may be interrupted; air pollution may occur	Keep sufficient spares always ready
AR brick-lined mixer (reactor) with paddles of special alloy	Thorough mixing of ground rock and acid	Production rate can get affected due to stoppage of the reactor	Keep ready spare shaft with paddles
Den Cutter for reacted mass	The cutters are used for the further mixing and reaction	Incomplete reaction	Keep sufficient spares ready always
Acidic liquor (H_2SiF_6) recycle system from scrubbers	H_2SiF_6 is recycled to mixer	If H_2SiF_6 recycle unit fails, the consumption of fresh H_2SO_4 will increase	Keep spare transfer pump for H_2SiF_6 recycle
Dilution of sulphuric acid and sending to feed tank for mixer (reactor)	For reaction with rock phosphate	If the dilution unit or transfer pump fails, the production rate is affected	Keep sufficient spares ready always
Air pollution control APC units (cyclone, venturi scrubber, tangential spray scrubber, spray tower, ID fans, chimney)	This is required to absorb the gases evolved from reactor and den for air pollution control	HF will be released into atmosphere and cause air pollution	Monitor operation of APC units and exit gases all the time
Precipitated silica removal from APC units	For smooth APC operation	System choking may occur	Must remove silica regularly/every day
Na_2SiF_6 production unit, effluent treatment plant	By product sodium silico fluoride for sale	Dilute H_2SiF_6 disposal problem	Na_2SiF_6 can give extra income

react with pure water (in the dilute NaOH solution) to produce NaOH and hydrogen. Chlorine is evolved in anode chamber. Both gases are evacuated separately from the cell. Depleted brine from the anode chamber is recycled for dissolving more salt. NaOH solution at higher concentration comes out from the cathode chamber. Rectifiers are used to supply high amperage direct current to the cells. Chlorine is cooled, dried by concentrated sulphuric acid, compressed, liquefied, and filled in cylinders. Likewise, hydrogen is also compressed and filled in cylinders. Both of these gases are then sold. They can also be used to produce HCl by burning Cl_2 in excess of H_2, then absorbing in DM water to produce HCl acid.

Any unreacted and unabsorbed gas is taken care by alkali scrubber (Fig. 16.2; Table 16.5).

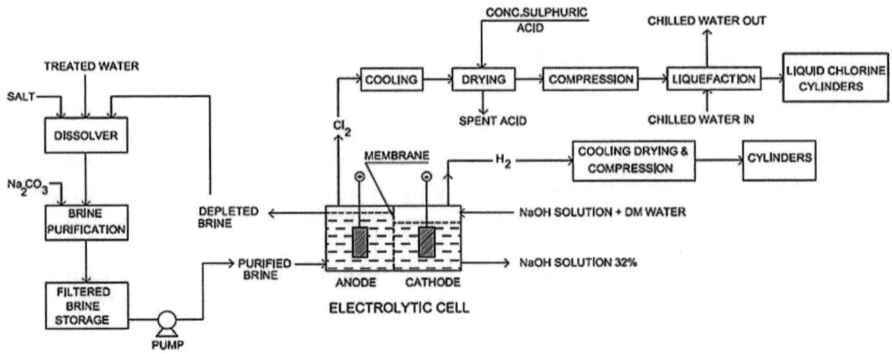

Fig. 16.2 Production of caustic soda and chlorine by electrolysis

Table 16.5 Checklist for manufacture of caustic soda and chlorine by electrolysis

Main equipments	Required for	Consequence for breakdown	Remark
Salt feeder and dissolver (fresh salt added to recycled spent brine along with treated water)	To make up brine strength for electrolysis	NaOH production can reduce if feed brine is diluted	Install more than one dissolver
Brine purifying units (addition of NaOH, Na_2CO_3, sodium bisulphate, and filters)	To have pure brine so that impurities will not choke membranes	Membranes may get blocked if the impurities enter the electrolytic cell	Monitor purified brine quality **on line**
Storage tanks for purified brine	This is required for continuous process	Uniform production rate is affected if storage of brine is not sufficient	Capacity shall be enough for 16–24 h requirement
Brine feeding system, flowmetres	This is required for feeding the brine into electrolytic cell	Electrolysis is affected if brine pump fails/flow is not proper	Keep spare pump, or gravity flow from storage
Electrolytic cells with accessories (anodes, and cathodes assemblies)	Required for electrolysis of brine	Production reduces if the electrodes do not work	Proper upkeep of electrodes and bus bars is necessary
Membranes and arrangements for fixing them	Required for diffusion of only Na^+ ions through membrane	NaOH production can reduce if the membranes are blocked with the impurities	Maintain brine purity; keep spare membranes
Rectifiers	Required for the conversion of AC to DC for electrolysis	Production stops if rectifier fails	Connect groups of cells to separate rectifiers
Gas evacuation from electrolytic cell	To prevent leaking out of gases from electrolytic cell or ducts	Dangerous situation may occur due to escape of Cl_2	Install Cl_2 gas detector; and gas pressure alarms

(continued)

Table 16.5 (continued)

Main equipments	Required for	Consequence for breakdown	Remark
Chlorine compressor, dryers, liquefaction system, chillers	This is required for drying and liquefying chlorine	Problem to handle excess chlorine; NaOH production may have to be reduced	Keep 200 % standby units and consume excess Cl_2 elsewhere
Chlorine cylinders filling arrangement	Liquefied chlorine is stored in cylinders	Serious safety issues in case of Cl_2 leak	Install Cl_2 detector and gas leak alarms
Hydrogen gas compressors, dryers and filling in cylinders	Hydrogen is stored in cylinders at high pressure	Serious fire/explosion issues in case of leak	Shall be no naked flame in plant; provide only flameproof fittings
Corrosion-resistant graphite furnace for HCl production	For burning of Cl_2 in H_2	Production of HCL will be curtailed	Cl_2 and H_2 gases to be disposed by other means
HCl gas absorbers and storage tanks	HCl gas is absorbed and stored in the tanks as HCl acid	If gas absorption system does not work, the HCl will be released as it is	Keep 200 % spares for absorber
Demineralised (DM) water plant	Required for the reaction with Na^+ ions to form NaOH	NaOH production curtailed	Keep sufficient DM water in storage tanks
Air pollution control (APC) alkali absorber Induced draught fans for areas of chlorine handling area, near electrolytic cells	This is required for the absorption of free chlorine/HCl and for removal of any Cl_2 leak	HCl gas will be released in the atmosphere; working area becomes suffocating and dangerous if Cl_2 is not removed	Keep gas detectors for Cl_2, HCl; 200 % spares for absorber and ID fans
Diesel generator (DG) sets for emergency power supply	Required for APC, lighting, instruments, HCl absorbers, Cl_2 liquefaction	Safety and environmental issues	DG set shall switch on automatically if grid power fails

16.6 Process for Manufacture of Nitric Acid by Oxidation of Ammonia

Liquid ammonia is evaporated by low-pressure steam and fed to the catalytic reactor at a controlled rate. Compressed air is filtered and preheated and is also fed to the reactor. Flow rates of air and ammonia are controlled to get about 10.5–11.0 % ammonia concentration. This is checked after the mixing unit (*before feeding to the reactor which operates at 850–900* °C). An arrester is provided to minimise the loss of platinised catalyst. The heat evolved due to burning of ammonia is recovered as steam by the waste heat recovery boiler (WHRB). Exit gases from WHRB preheat the incoming air. The NH_3 is oxidised to nitric oxide (NO), which is then further oxidised to NO_2 and then absorbed to produce nitric acid. Some NO is generated again which is again oxidised to NO_2 by air injection in

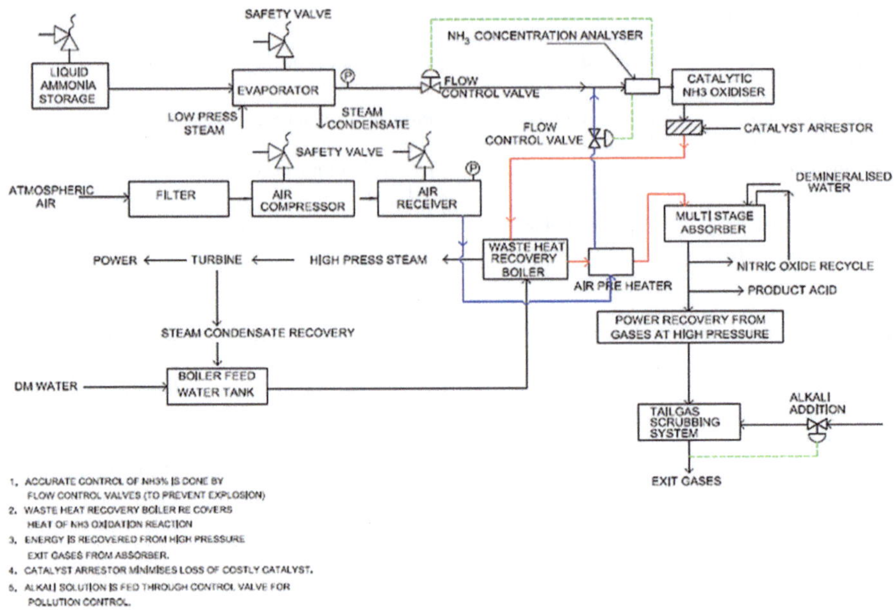

Fig. 16.3 Manufacture of nitric acid by oxidation of ammonia

the absorber. Gases come out from the absorber at pressure, and power recovery is done through special turbines. Alkali solution is added automatically to the tail gas scrubbing system for pollution control.

Green lines shown in Fig. 16.3 indicate signals for analysers for process control.

Main equipments	Required for	Consequence for breakdown	Remark
Liquid ammonia storage	Ammonia is the main raw material which is required for formation of NO initially and NO_2 further on	Safety issue if ammonia leaks or if hot ammonia from evaporator flows back into main storage	Provide safety valves on NH_3 storage, and NRV on liquid outlet towards the evaporator
Evaporator	Liquid ammonia is vapourised by using low-pressure steam as a heating media	Safety issue if it is overpressurised due to high- pressure steam and ammonia leak occurs	PRV on steam line and safety valve to prevent overheating
PRV—pressure regulating valve	Steam pressure regulation	Overheating of evaporator	Provide safety valve on main steam line also
Flow control valve and pressure gauge	Control of ammonia flow for optimum concentration in feed gas	Safety issues if ammonia concentration in feed gas is high	Online gas analyser to be provided

(continued)

(continued)

Main equipments	Required for	Consequence for breakdown	Remark
Air compressor unit with filtration and receiver	Filtered air is required for oxidation of vapourised ammonia	If filtration system does not work the impurities may disturb control valve operation and also the quality of product	Air receiver must be have safety valve
Catalytic NH_3 oxidiser with platinum-based catalyst	This is necessary to achieve high oxidation rate	Escape of unreacted ammonia if catalyst malfunctioning occurs	Catalyst arrester is necessary to minimise the loss of catalyst
Waste heat recovery boiler (WHRB)	Heat generated during oxidation of ammonia is recovered to generate steam	If water flow gets interrupted the boiler will be in danger	Water-level controller is a must; stop gas feed if water level is too low
Demineralised (DM) water plant	DM water is required for the absorption of NO_2 to form HNO_3 and also for WHRB	Will affect the quality of product and operation of WHRB	Use of untreated water not advisable
Multistage absorber	NO_2 is absorbed into DM water to achieve the maximum yield	If the multistage system fails, production of acid will be less	Install online analyser for NO and NO_2
Start up boiler with accessories, fuel storage, air blower	For initial heating of liquid NH_3 by steam for plant start-up	Other means required to vapourise liquid ammonia for starting the plant	Use exhaust steam if available from turbogenerator
Power recovery system	High-pressure gases can be used to generate power	Plant efficiency decreases, operating pressure settings need adjustment	Turbine shall be compatible with corrosive gases

16.7 Steam Generation by Coal-Fired Boiler

Main equipments	Required for	Consequences of breakdown	Remark
Coal pulveriser	To pulverise coal	Big pieces of coal will be conveyed	Provide grid to reject very big pieces
Vibratory screen for coal	To separate bigger pieces	Big pieces may get stuck in burner	Keep spare screen, bearings, motor
Belt conveyor	To fill up feed hopper	Firing may get interrupted in furnace	Feed hopper capacity shall be at least for 4 h
Screw feeder	To feed coal into burner	Firing *will get* interrupted in furnace	Provide oil burner as standby

(continued)

(continued)

Main equipments	Required for	Consequences of breakdown	Remark
Forced draught fan	To supply combustion air	Firing *will get* interrupted in furnace	Provide one more fan as standby
Boiler feed water (BFW) tank	To supply treated water to boiler	Boiler will be in danger	Provide low-level alarm to the tank
Boiler feed water pump	As above	Boiler will be in danger	Provide one more pump as standby
Non-return valve in BFW line	To prevent backflow of steam into BFW line	Steam backflow can damage BFW pump	Shall be of standard approved design and make
Blowdown valve	To drain out water from boiler with high concentration of dissolved solids	Deposits of salts may occur on tube affecting efficiency of heat transfer and life of tubes	Regular blowdown and water analysis is necessary
Forced draught fan	To suck out flue gases	Firing *will get* interrupted in furnace	Provide one more fan as standby
Economiser/air preheater	To increase thermal efficiency	Fuel consumption will increase	Keep these units well maintained
Ash separators	For control of pollution caused by fly ash	Environmental pollution will occur	Provide electrostatic separators
Mountings and fittings Safety valves, gauge glasses, level controllers, steam stop valves	For smooth operation and to meet statutory regulations	Steam supply to plant will be interrupted; permission to operate will be withdrawn	Comply with all statutory regulations, low water-level alarms and trips for very low water level
Thermal insulation on outside and refractory lining inside	To increase the thermal efficiency	Fuel consumption will increase	Keep it well maintained

16.8 Some Useful Formats for the Production Managers

The formats and items listed in them are only indicative and plant managers/senior engineers shall revise them to suit their own plants.

16.8.1 Production Record of a Plant

Name of Product- .... month and year..........

Date	Production on date MT	Progressive production up to date MT	Target production or (rated production) MT		Difference on date	Prog. difference up to date	Remarks
			On date	Up to date			

16.8.2 Record of Consumption of Raw Materials

Name of Raw material............ month and year.........

Date	Consumption on date	Progressive consumption up to date	Consumption as per standard norms		Deviation from norms		Remarks
			On date	Up to date	On date	Up to date	

The remarks column can include information about the vendor and analysis of raw material:

Similar records shall be maintained for recording consumptions of power, water, fuels, and packing items.

16.8.3 Record of Expenses Incurred on Maintenance of the Plant

Name of equipment.......—location in plant /Tag no ......

Date	Name of part	Job done	Cost of spare used	Stoppage of plant	Remarks

16.8.4 Record of Raw Material Stock Position

Name of material………….. Item Code………………..

Date	Opening stock (MT)	Receipts (MT) progressive	Issues (MT)	Closing stock	Closing stock will last for (days)

16.8.5 Record of Sale of Products

Name of Product Code

Date	Opening stock	Saleable product added to stock	Product sold from stock	Closing stock of saleable product	Stock of product not as per specs	Remark

16.8.6 Rejected and Returned Products

The products not meeting specifications could be purified, recycled to plant, or sold at lower price to parties who can accept it for some other use.

A separate record shall be made for any product returned by the customers. Reasons such as not meeting certain specifications (less purity, the presence of insoluble matter, high moisture, material not free-flowing type, etc.) shall be clear so that such matters can be suitably addressed in future consignments. Complete history of production of such items shall be traced out and analysed (date of production, batch no or reactors used, raw materials used and their analysis, operating conditions since two days before, and on the date of production itself, any abnormality noticed on that particular date.).

16.8.7 Record of Stores Inventory

Name of Spare……Required for………. Code number….. Re-order point

Date	Opening stock	New items procured	Closing stock	Closing stock will last for (days/weeks)	Details of new order	Remark

Reorder point is when new orders must be placed.

Lists of important spares
Make a list of spares which contribute about 80 % of the total cost. This indicates a considerable share of the amount spent on maintenance. Also make another list of the more frequently required items. Both these lists can help draw attention to the right areas. They shall be looked into the reduced cost of maintenance and frequency of replacements.

There shall be a special budget for items for personal safety and ETP and for expenses required to comply with instructions from statutory inspectors (*work on pressure vessels, electrical equipments, boilers*).

Chapter 17
Identifications of Hazards: Some Practical Examples

The following tables give initial identification of hazards after study of the process flow diagrams (PFD) and basic engineering documents (BED). These tables are only for illustration purposes.

17.1 Identification of Potential Sources of Hazards

In actual practice, the plant engineers shall carefully study the PFD for their own new projects (or existing plants if expansion/diversification is being planned) and find out potential sources of hazards (HAZID), probabilities of occurrence of harmful events, and their likely consequences such as injuries to persons, environmental pollution, damage to plant equipments, and loss of production.

Their observations and apprehensions shall be conveyed to the designers (*with requests for addressing the issues raised by making appropriate changes*). The designers shall send back the revised documents when the design is complete with:

- Details of the process and required equipments,
- Plant layout,
- Material and energy balance,
- Piping and instrumentation diagrams.
  Operating procedures may also be examined to ensure that all eventualities have been considered.

The full study procedure of the HAZOP study shall be carried out **now** before proceeding with erection and commissioning. This will avoid costly changes in the plant later on and will also help to assess the operability of the plant.

The study will be found very useful for the existing plant and new equipment (when expansion/diversification will be planned). Therefore, a project may be studied several times in its lifetime.

K.R. Golwalkar, *Production Management of Chemical Industries*,
DOI 10.1007/978-3-319-28253-4_17

17.2 Case Study: Manufacture of Sulphuric Acid

It was proposed to start a new sulphuric acid plant. The DCDA process was chosen with elemental sulphur as raw material. The basic engineering document giving the process flow sheet was submitted by the consultants for comments by senior plant engineers. Certain hazards were identified and a report was sent to the consultants for modifying the technology. The issues raised by the HAZID studies and the suggestions were looked in. The designs were modified and detailed P and IDs were made. These were studied again and request was made for some changes so as to make the plant safe for operation. The suggestions were addressed while finalising the detailed engineering designs, drawings, and equipment specifications.

The process for the manufacture is given here in brief for ready reference. Solid sulphur is melted by steam coils and pumped to the furnace for producing SO_2-bearing gases. The heat released due to burning of sulphur is recovered by waste heat recovery boiler as steam (which is used to melt the sulphur). The gases are passed through a multistage converter for conversion of SO_2 to SO_3 which is absorbed in interpass and final absorption towers in circulating streams of acid at optimum temperature and concentration. Provision of candle demister and alkali scrubber is made after final absorption tower to prevent the environmental pollution. An air blower provides air required for the combustion furnace. The air is dried in the drying tower by stream of acid at optimum temperature.

In Table 17.1, only the identification of hazards has been shown. The P and IDs are proprietary and hence not shown.

Table 17.1 Identification of some hazards in sulphuric acid manufacturing plant

Identification of hazards in plant	Consequences	Preventive steps	Remark/requests to designers
Sudden leak of steam coil in melter	Burn injury to personnel	Provide safety valve on the steam line	[a]See below
Fire at sulphur yard or melters.	Plant area and surroundings will be heavily polluted by SO_2 gas	Do not allow smoking in this area; provision of fire fighting system	Ban smoking strictly; check operation of fire fighting system regularly
Sulphur feeding continued even when air supply is stopped	Accumulation in furnace may cause explosive conditions; carryover of un burnt sulphur may cause problems in downstream units	Sulphur pump should stop when air blower stops;	Check interlock with blower at least once every week
Low water level in waste	Chance of leak of water tubes of boiler	Automatic level controller and audio	Air blower and sulphur pump should trip at **very** low level of water

(continued)

Table 17.1 (continued)

Identification of hazards in plant	Consequences	Preventive steps	Remark/requests to designers
heat recovery boiler WHRB		visual alarm at low level provided	
Improper conversion of SO_2 in converter	Environmental pollution due to unconverted SO_2	Run alkali scrubber after final absorption tower continuously	Online monitoring of SO_2 in exit gases with alarm
Improper absorption of SO_3 and acid mist in absorption tower	Unabsorbed SO_3 and acid mist in exit gases	Provision of candle demister and alkali scrubber after final absorption tower	Online monitoring of SO_3 and acid mist in exit gases with alarm
Acid pipe or flanged joint leak	Burn injury to personnel	Provide covers on all flanged joints, valves	See below
Storage tank leak	Burn injury to personnel	Provide dyke walls, keep one storage tank always empty	Provide transfer pump for acid collected in dyke walled enclosure
Gas line leak	Environmental pollution	Select the pipes for high-temperature service; provide expansion bellows; put covers on pipes	Use pipes of sufficient thickness, carry out stress analysis, provide sliding supports if required
Less flow of scrubbing liquor/low level in tail gas scrubber tank	Environmental pollution may occur	Low flow alarm; test level indicator alarm every week	Provide automatic addition of alkali to scrubbing liquor tank

Following changes were incorporated in plant design after a study of the above:
[a]**Steam supply to melters**:
Pressure reducing valve on steam supply set at 5 kg/cm^2
Provision of an additional safety valve on main supply line and covers on melter
Application of some typical guide words such as no flow, reverse flow, excess flow was made to further improve the plant design

17.2.1 Identification of Some Hazards in Sulphuric Acid Manufacturing Plan

Sulphur feed:

Excess feed/Less feed: Provision of metering pumps for controlled sulphur feed to prevent excess or less feed.

Excess feed can happen with centrifugal pumps hence provide variable-frequency drive motor and low/high SO_2 % alarms for gas composition; and furnace exit gas temperature *if it is not possible to provide metering pumps.*

Reverse feed
A reverse flow of sulphur from the combustion furnace to the melters cannot happen through the spray gun even when the sulphur pumps trip (in this case, temperatures will drop in all units. It is indicated by alarm for sulphur pump and by temperature indicators).

Boiler feed water

Reverse flow—Non-return valves are to be provided in boiler feed waterline to prevent reverse flow of steam/water from boiler into the pumps.

Less/no flow of feed water to boiler—Provision of electrical interlocks for tripping of air blower (and consequently the sulphur feed pump also) when water level is very low in WHRB when there is *very less* or *no flow* of feed water to the boiler.

Acid flow to towers (for drying of air and for absorption of SO_3)

Excess feed—Carryover of acid particles along with gas flow can occur. Hence, the acid-exit nozzles of all towers shall be of at least twice the diameter of acid inlet nozzles to prevent accumulation inside. Demisters shall be provided to prevent carryover of acid particles along with gas flow if excess acid flows in.

Reverse flow—Not possible from towers to acid supply pumps.

Less/no flow of acid to towers—Can cause environmental pollution; hence provide alarm on acid flowmeter if the acid flow is less or there is no flow. Monitor flow of acid by current drawn by circulation pump motor also.

[Provide partitions to cordon off area near acid lines and coolers; provide eyewash fountains and safety showers].

Airflow to furnace

Reverse flow The plant operates at a pressure a little above atmospheric (1500–2500 mm WG generally). Provide a **non-return valve** in the air blower discharge line to prevent the reverse flow of air from plant units to air blower *in case the blower trips*. This will also be useful if more than one air blower is run together to supply air.

Less/no flow of air to plant units (drying tower)—This can cause loss of production, improper combustion of sulphur in furnace. Hence, it provides an air flowmeter, pressure gauge/manometer in airline. Operate the plant with positive displacement-type air blower.

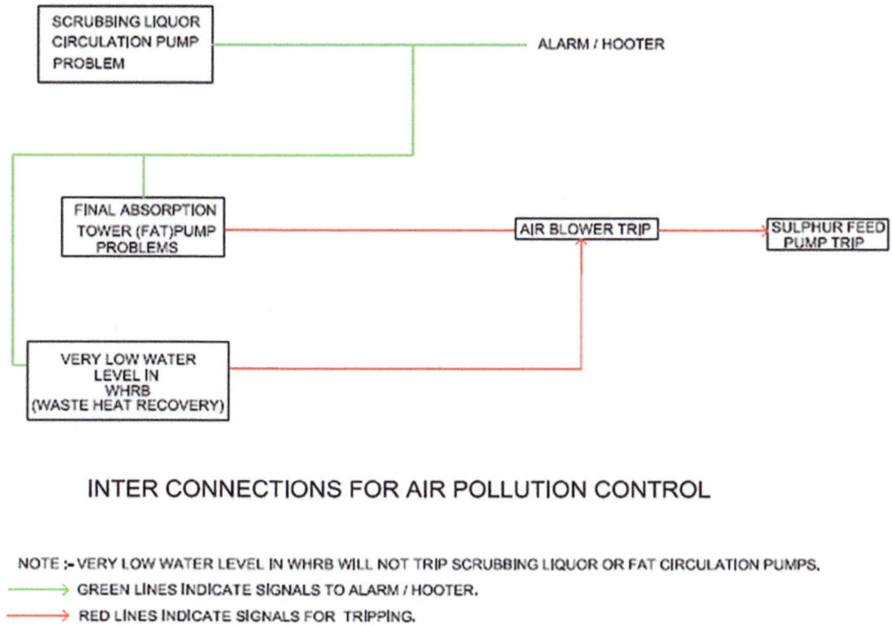

Scrubbing liquor tank

Low level—Environmental pollution may occur if the level goes down below suction nozzle of the pump and it does not deliver alkali liquor to the scrubbing tower.

Provision of automatic level controller and low-level alarm for the tank suggested.

High-level liquor may overflow with the loss of alkali. Provide float-operated water addition valves for level control of the tank.

High and low temperatures in plant units (converter gas inlet and exit).

Provide audio visual alarms in control room for drawing attention of operators towards such events. These are likely to disturb the conversion of SO_2 to SO_3. Strict instructions shall be given in operating manuals. Alkali scrubbers for tail gases shall be kept running always with proper alkali make-up.

17.3 Illustration: Manufacture of Nitric Acid

The process is given in Chap. 16. Table 17.2 is the identification of some hazards in this plant. The list is not exhaustive and more identified hazards should be added for modifying the process, equipment, and operation to make the plant safer.

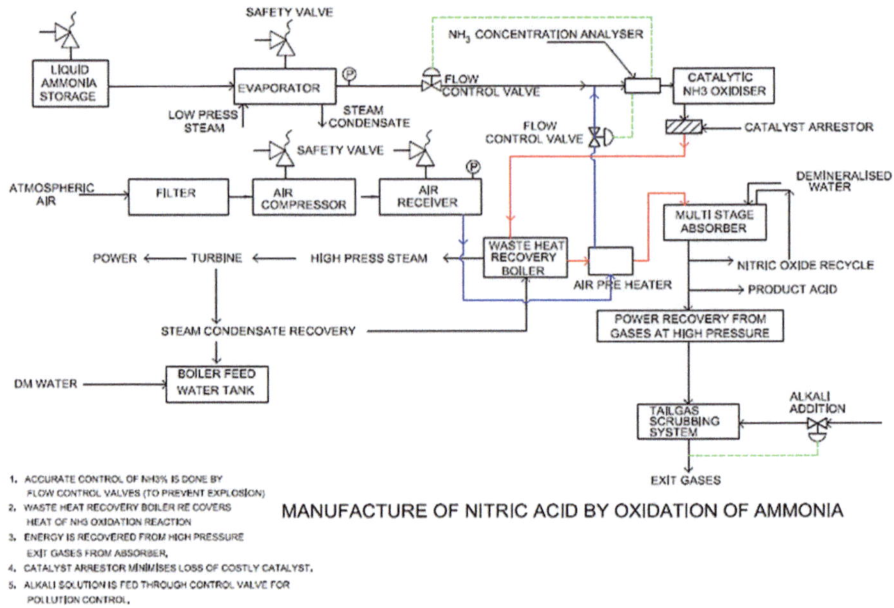

17.3.1 Identification of Some Hazards in Nitric Acid Manufacturing Plant

See Table 17.2.

Table 17.2 Identification of some hazards in nitric acid manufacturing plant

Identification of hazards in plant	Consequences	Preventive steps	Remark/requests to designers
Ammonia storage tank leak	Injurious to personnel and environment	Design and construction of storage system must be as per statutory regulations	Provide safety enclosures and devices as per international standards
Overheating by steam heater; leak of ammonia pipes and valves	Unit may leak suddenly	Design and construction of the system must be as per statutory regulations	Provide pipes, valves, flanges, etc. safety devices as per international standards

(continued)

Table 17.2 (continued)

Identification of hazards in plant	Consequences	Preventive steps	Remark/requests to designers
Chance of explosion in converter	Major accident	Provide safety valves and rupture discs control NH_3 % in gas mixture accurately	Design and construction of the system must be as per international standards; check control valves for air and ammonia regularly
Less supply of demineralised DM feed water to WHRB	Danger to boiler;	Standby BFW pumps; and storage of DM water	Sufficiently large separate tanks for WHRB and absorption tower
Less supply of air to converter	Chance of explosion in converter	Alarm for high NH_3 concentration in gas mixture; reduce/stop ammonia supply if air supply is less	Regular gas analysis; proper maintenance of feed control valves; standby air compressor
Less supply of demineralised DM feed water to absorption tower	less absorption of NO_2 can cause environmental pollution; the tower can get overheated and over pressurised	Standby feed water pumps and sufficient storage of DM water provide safety valves for pressure release	Design and construction of the tower as a pressure vessel; provide safety valves for pressure release
Chloride content in DM water	Can corrode absorption tower and pipes, and spoil product quality	Strict check on DM water analysis	Provide online conductivity metre for DM water; carry out regular manual checks
APH air preheater shell leaks (WHRB exit gases on shell side)	Pollution in plant area	APH construction as per ASME Sec VIII and statutory regulations	Regular inspection; shell thickness testing, pressure tests must be done.
APH air preheater tubes leaks (compressed air on tube side)	Air supply to converter reduces and NH_3 % may increase dangerously in converter inlet gas	APH construction as per ASME Sec VIII, IX and statutory regulations, change tubes every two years	Regular gas analysis at converter inlet; proper maintenance of feed control valves
APH air preheater shell leaks (compressed air on shell side)	Only air will come out		NH_3 % increase too much in converter inlet gas only if air leak is major
APH air preheater tubes leak (WHRB exit gases on tube side)	NH_3 % increases too much in converter inlet gas only if air leak is major	APH construction as per ASME Sec VIII, IX, and statutory regulations, change tubes every two years	Regular inspection; shell thickness testing, pressure tests must be done

(continued)

Table 17.2 (continued)

Identification of hazards in plant	Consequences	Preventive steps	Remark/requests to designers
Malfunctioning of ammonia and airflow control valves	Chance of explosion in converter	NH_3 % analyser at converter inlet; provision of manual override	Consider additional stop valve in NH_3 exit line from evaporator to converter
Exit gases from absorber contain nitrogen oxides, HNO_3 mist	Environmental pollution	Exit gas analyser and alkali scrubber for exit gases	
Leak from nitric acid lines	Injury to personnel	Use stainless steel pipes of high-pressure rating; covers on all flanged joints	Design and construction as per high-pressure piping system.
Storage tank leak	Injury to personnel	Provide dyke walls, keep one storage tank always empty	Provide transfer pump for acid collected in dyke-walled enclosure

17.3.2 Steam Supply to Ammonia Evaporators

- Pressure reducing valve on steam supply set at 5 kg/cm^2.
- Provision of an additional safety valve on main supply line.

17.3.3 Application of Some Typical Guide Words

Application of some typical guide words such as no flow, reverse flow, excess flow, etc. is to be done for ammonia and airflow, DM water supply to boiler and absorber after the process design, and equipment details and P and ID are completed. This shall help to further improve the plant design.

17.4 Manufacture of Caustic Soda (Membrane Process)

The process is given in Chap. 16.

17.4.1 Identification of Some Hazards in Caustic Soda Manufacturing Plant

Some further requests can be sent to the designers as below (Table 17.3).

<u>Flow of fresh brine</u>

Excess flow and less flow do not pose safety hazards.
No flow—will interrupt production.

Table 17.3 Identification of some hazards in caustic soda manufacturing plant

Identification of hazards in plant	Consequences	Preventive steps	Remark/requests to designers
Cl_2 gas leak from anode side of electrolytic cell	Can be very dangerous if inhaled by personnel	ID fans for working area; plant operation on chlorine side under suction, provision of gas masks	Provide spare suction fans and fresh ventilation for working areas
Electrical shock	Can be fatal	Provide covers on all open wiring, bus bars, rubber sheets on floor	Persons must wear shoes with thick rubber soles
Leakage of chlorine from drying, compression, or liquefaction system	Can be very dangerous if inhaled by personnel	Install gas detectors and alarms for Cl_2 in working areas provision of gas masks	Provide spare suction fans and fresh ventilation for working areas
Acid leak from chlorine-drying system	Severe burns to personnel due to sulphuric acid (fresh and spent acid)	Full covers on acid lines and pump discharge; safe disposal of spent acid necessary	Safety showers, eyewash fountains in working areas, use personal safety devices
Hydrogen leak	Fire/explosion hazard	Avoid naked flames, electrical loose connections, appropriate vents at high spots	Provide spare suction fans and fresh ventilation for working areas
Escape of Cl_2 or HCl from tail gas scrubber	Can be dangerous	Provide alkali scrubbers, exit gas analysers, and gas detectors in working areas	Provide installed spare circulation pumps for scrubber; install gas detectors for Cl_2 and HCl and alarms
Leak from caustic pipe lines	Severe burns to personnel	Provide covers on all flanged joints	Safety showers, eyewash fountains in working areas, use personal safety devices

Ventilation

Less suction and complete stoppage of suction fans can be very dangerous as any small quantity of chlorine leak can be inhaled by persons in the working area. Provide installed standby suction fans.

Tail gas scrubbers

Low flow/no flow of scrubbing liquor can result in the release of unabsorbed Cl_2 or HCl. Provide installed spare alkali circulation pumps for scrubber. Install gas detectors at exit of scrubbers for Cl_2 and HCl and provide alarms for warning the operators.

Chlorine drying system; caustic soda solution handling area

Any leak can cause burn injuries (due to sulphuric acid/caustic soda solution). Provide covers on all pumps, flanged joints of pipelines, and provide safety showers, eyewash fountains in working areas.

Chlorine liquefaction

In case of low flow/no flow of brine, there could be high pressure of Cl_2 at the condenser inlet. Provide alarm if the brine flow is less, and either divert Cl_2 for HCl production or stop electrolysis to stop generation of more Cl_2.

Provide installed spare brine circulation pumps for Cl_2 condenser.

17.4.2 Application of Some Typical Guide Words

Application of some typical guide words such as no flow, reverse flow, excess flow is to be done for brine flow, water supply to the plant, chlorine handling system, HCl scrubber, etc. after the process design, and equipment details and P and ID are completed. This shall help to further improve the plant design.

Chapter 18
General Technical Matters

18.1 Batch and Continuous Production

18.1.1 Salient Features of Batch Production

This chapter focuses on the manufacture of different types of similar products in batches (e.g. different organic chemicals, pharmaceutical items, steam distillation of different grasses and plant material, dyes, and types of paint). Generally, a single production line can be used for many products, and hence, it can reduce initial capital required for setting up many production units.

- Can be suitable when the raw materials are very costly or the finished product is a special item and can be sold at a high price.
- Specific requirements of the customer can be met by adjusting compositions and processing methods. The product may be tailored for a particular customer.
- Batch production is also useful for seasonal items, products for which it is difficult to forecast demand, for a trial run for production, or products that have a high-profit margin.
- **Suitability for Research and Development**—Quality of product can be changed for different batches, and hence, effect of change in composition or operating conditions can be studied over a small volume of product.
- Generally, small quantities are produced at a time. The batch size is to be decided in consultation with the sales (marketing) department. Only a small quantity of off-spec product gets produced if something goes wrong. Batch production is thus suitable if any batch does not meet specifications or if any consignment is rejected by the clients.
- It is possible to trace out the raw materials/ingredients actually used, process conditions, or operational steps carried out (**or** *mistakes made can be analysed easily*) if all batch numbers are recorded properly. Necessary corrections can then be made for the next batch.

K.R. Golwalkar, *Production Management of Chemical Industries*,
DOI 10.1007/978-3-319-28253-4_18

In case of a continuous production plant, a large quantity of off-spec product may get produced by the time the change in inputs results in better quality. This can be a big loss if this material cannot be reprocessed to meet the required specs.

- The raw materials and other inputs need not be consumed if there is not enough demand for the products. These items can be kept in stock till sales department confirms that demand for the product has picked up.
  This is generally not possible for a continuous production plant and a certain minimum quantity of inputs gets consumed corresponding to the lowest rate at which the production units have to be run as per design.
- Batch production can therefore be useful for small businesses who wish to minimise the initial investments and working capital required for the *continuous production lines*. The producer can cease production any time without having to keep the plant running (which can result in losses).
- The work force should have labour having more skills—who can carry out all the steps of a batch—which is necessary because the complete product is made in a batch, as compared to a continuous production line having many work stations—wherein only a limited type of work is generally done at each of them.
- Batch production can be kept stopped as per convenience and hence work force need not be engaged in all shifts. This can save considerable costs as relievers are not required on weekly off days.
- Batch production may need resetting of process units after every batch or a complete stoppage and thorough cleaning of all process units used (*to avoid contamination by the traces of material processed in previous batch*) if the same equipment is used.

18.1.2 Cycle Time

Cycle time is the time between two consecutive batches of identical product. Only resetting of process units may be required for the next batch.

18.1.3 Idle Time

Idle time is the *changeover time* between batches of different products since many equipment is not being used for production till the cleaning is over and preparations such as lifting up bags of raw materials to the charging floor, resetting of process units are going on to start the next batch. The resources are idle during this time.

Consequently, production managers have to plan the production schedules to minimise changeovers. This shall be done in consultation with the marketing (sales) department.

18.1.4 Maintenance of the Process Units

Maintenance of the process units can be taken up during the changeover time in many batch production plants. It is usually not necessary to stop the production since sufficient time may be available. This is difficult in case of a continuously running plant where interruption of the production generally takes place when the maintenance activities are to be carried out.

18.1.5 Salient Features of Continuous Production

- Continuous production requires special machinery and higher capital investment. It is essential to achieve high level of utilisation and the management is therefore very particular about not losing any productive time.
- This requires a high level of sales of a fairly standardised product made for the market. However, unit costs will be lower as compared to batch production due to the high-volume production runs.
- Production workers in continuous production commonly work in shift duties and hence need relievers. In case a reliever does not report on duty due to any reason, the person from previous shift cannot leave the work station and may have to be engaged on overtime basis till another person can relieve. *This can become a critical issue if the plant units are pressure vessels, or carry dangerous materials.*
- There are sequential shutdown and start-up procedures for complex continuous plants that must be carefully followed for safety of plant and personnel. Typically a start-up or shutdown will take several hours. After starting, a complex continuous process plant may need automatic controllers to control the flow rates, concentrations, pressures, temperatures, levels in circulation and overhead tanks, etc.
- The quality of products during shutting down and starting-up of many continuous processes plants may not be as per specifications. (*since adjustments of process controls are required during starting and stopping.*)
  These off-spec products will have to be recycled, purified, sold at lower price, or disposed of through effluent treatment units/incinerators.
- Process units (reactors, condensers, and absorption towers), pipelines, and intermediate storages will have to be completely drained out and cleaned whenever the plant is shut down to prevent corrosion and choking. There could be settling of material or even formation of hard sludge
- The units lined with refractory (kilns, blast furnaces, and converters) can get thermal shocks due to frequent heating and cooling because of starting-up and shutting down. It is not advisable to frequently start and stop certain pressure

vessels because these operations may be done at a fast rate rather than following the manufacturer's instructions every time.
- These in turn may cause metal fatigue or other wear from pressure or thermal cycling.

18.2 Pressure Vessels

These are some of the most important equipment in a chemical industry.

Typical applications are manufacture of ammonia, liquid sulphur trioxide (by boiling of oleums), nitrogen and oxygen (by Cryogenic, PSA, and membrane processes), air receivers for compressed air systems for process control, waste heat recovery boilers, etc. *The production management should treat any process unit which is likely to develop pressure due to any reason (such as excessive heating, or any reaction going on inside—e.g. an agitated tank process reactor) as a pressure vessel from safety point of view. It can be a condenser for ammonia or SO_2 also is it can get pressurised if the cooling water supply fails.*

18.2.1 Duty Conditions to Be Considered

- Intended use as a reactor, for storing liquefied gases, and for boiling process liquids
- Volumetric capacity **of** the vessel
- Whether used with internal agitator or by sparging gases inside
- Will it be a fired or an unfired vessel? This is very important for statutory regulations.
- Properties and composition of material to be handled—*special care required for toxic, inflammable, and dangerous materials*
- Operating pressures and temperatures—normal working and maximum
- *Possibility of any sudden rise in temperature and/or pressure due to sudden (uncontrolled) higher rates of reactions must be considered. This can happen if excess of reactants are fed in, cooling system fails, use of higher concentration of reactants,* etc.
- Installation position and location—horizontal or vertical; indoor or outdoor (exposed to sun, rain, and snow fall)
- Continuously or intermittently pressurised during use (air receivers, nitrogen surge tanks, etc.)—this can result in mechanical fatigue
- Is the vessel installed at a fixed location in the plant or to be used for transportation (LPG or oxygen tanker) or to be used during travelling compressed refrigerant systems for trucks, in air-conditioned passenger trains and buses

- Arrangement for heating/cooling—*Whether by external limpet coils* (*these give external reinforcement*) *or internal coils* (*these are difficult to clean and replace*)
- Connecting pipes can cause bending loads or mechanical stresses on the pressure vessels if they are not supported independently outside, and can shorten their life.

18.2.2 General Arrangement Drawing

- Mention main duty conditions
- Nozzles required for level indicator, sight and light glasses, sampling points, inlet of reactants and exit of products, entry for agitator shaft, thermowells for temperature indicators, vent valves, and pressure and temperature indicator probes.
- Dimensions of connecting pipes for incoming and outgoing materials and draining out the vessel.
- Manholes for inspection and maintenance
- Safety valves and Rupture discs.
- *Details of pad plates and lifting lugs*
- Drain valves and vent valves
- Material of construction and welding electrodes to be used
- Cleats required for external insulation.

18.2.3 Finalise Detailed Arrangement

Discuss GA given by consultants with senior engineers and finalise all details shown.

18.2.4 Mechanical Design

Complete the mechanical design as per standard codes while keeping in mind the important temperature limitations of materials of constructions as follows:

Carbon steel for non-corrosive service—up to 500 °C
Stainless steels (*18–25 % Cr*; *8–20 % Ni 0.05–0.2 % C*) for use in corrosive conditions up to 600 °C.

18.2.4.1 Proceed to Prepare Fabrication Drawings

Mentioned in the drawing are the fabrication codes to be followed for construction. Some well-known codes are as follows:

- ASME Boiler and Pressure Vessel Code Section VIII: Rules for Construction of Pressure Vessels.
- ASME Sec IX to ensure the fabricator, welders posses the necessary qualifications for welding and brazing
- Pressure Equipment Directive (PED)
- Japanese Industrial Standard (JIS)
- Indian Boiler Regulations IBR 1950 *with latest amendments*
- German Pressure Vessel Codes (DIN).

IS 2825–1969 for Unfired Pressure Vessels (these are Indian codes. Equivalent international codes shall be used)

They classify the vessels as follows:

- **Class I Vessels** which carry lethal/toxic substances–Full radiography is mandatory. Structural steels such as *IS 226/2062 are not allowed for their construction.*
  Only IS 2002 Gr A (boiler quality steel) is allowed
- **Class II Vessels**—commonly used in chemical process industries. *All longitudinal and circumferential joints should be radiographed.*
- **Class III Vessels**—Used for light duties only (pressure not more than 7.5 kg/cm^2, temperature maximum 250 °C.)

18.2.4.2 Submit Fabrication Drawings for Statutory Approval

Obtain sanction with seal of approval for release for construction. Six sets of original documents shall be made—for own Master Record, for statutory authorities, for designer, for fabricator, for chief engineer, and for plant engineer.

18.2.4.3 Placing Order for Fabrication

Inquiries shall be floated for short listing, followed by technical and final commercial negotiations. Management shall place orders on the finally selected suitable fabricator after this.

18.2.4.4 Quality Assurance Plan

Quality assurance plan—to be obtained from vendor/fabricator.

- Vendor must possess valid certificates for fabrication of pressure vessels (as required by statutory authorities)
- Design and fabrication codes followed, safety margin considered, and Corrosion allowance added
- Traceability of material used (*original bills, test certificates of materials used,* etc.)
- Vendor must test all required mounting and fittings, bought out components as per statutory regulations
- Records of post-weld heat treatment done for stress relieving.

18.2.4.5 Non-destructive Tests

These are to be carried out in the presence of senior plant engineers deputed by plant management.

- Visual checks of all dimensions, nozzle orientations, fittings, etc.
- Ovality checks (refers to maximum diameter minus minimum diameter.)
  The pressure vessel may de-rated if the ovality is beyond acceptance limits given by designer. It may be used only at a lower pressure and temperature freshly calculated by the designer and finally approved by statutory inspectors.
- Dye penetration/magnetic particles tests—to detect cracks in welded joints
- Hydraulic pressure test (only chloride-free demineralised water shall be used),
- Calibration test records for the pressure gauges (used during NDT) shall be preserved and made available to purchaser as well as statutory authorities on demand.

Radiographic Testing

All records of such tests shall be preserved.
Radiography of welded joints shall be as per approved fabrication drawing.

18.2.4.6 Statutory Documents

Statutory documents—*These are required in many countries for obtaining permission to erect and hydro-test (and for future reference/records of the purchaser).*

- Documents for intended use/service conditions,
- GA and approved fabrication drawings with inlet and exit nozzles, their orientations, nozzles for safety valves, and vents.

18.2.4.7 Post-weld Heat Treatment

- PWHT is mandatory when vessel is >38 mm thick
- When lethal fluid is handled,
- Required temperature for PWHT shall be as per ASME Sec VIII guidelines.

This shall be insisted by the purchaser before accepting the pressure vessel because it can:

- *relieve stresses introduced during manufacture*
- *improve corrosion resistance*
- *improve strength and toughness*

No changes or corrections involving cutting/welding are permitted after PWHT.

18.2.4.8 Hydrostatic Test

The test shall be carried out as per procedure instructed by statutory authorities. Generally, the test pressure and the duration of the test are instructed. The tests are to be carried out in the presence of the statutory authorities.

18.2.4.9 Safety Valves

- Must be procured with proper test certificates only
  Setting of safety valves shall be done as directed by statutory authorities
- At least two safety valves are to be provided.
- Reinforce the nozzles on the pressure vessel for mounting the safety valves
- Distortion of valve body can occur due to the weight of blow-off pipe and over tightening of bolts. Hence, *proper supports and clamping required.*

Some Practical Tips

- Do not depend on one safety device alone. Provide at least two safety valves. Provide rupture disc in addition to safety valves if possible.
- Spring-loaded valves shall have protective covers on the springs so they do not get jammed due to corrosion, rains, dust, etc.
- Avoid a long pipeline for connecting safety valves to the vessel, since it can get choked due to the deposition of solids from inside.
- Bends should be avoided in the connecting pipeline.

- Counter weight must be fixed properly.
- Release port must be of adequate size.
- Keep the exit and inlet pipes of safety valves always clean.
- Corrosion of seat—continuous blow-off can lead to the loss of material from the vessel.

18.3 Plant Operation at Different Pressures

18.3.1 Plant Operation Below Atmospheric Pressure

Some chemical plants are operated at atmospheric pressure or even below.

- Suitable for toxic, poisonous, and corrosive gases. In case of a leak, no gas will come out from the leaking spot.
- Not suitable for inflammable or explosive vapours
- Need an induced draught (ID) fan which can create sufficiently strong suction throughout the plant (it should be able to overcome the pressure drops in all units and connecting ducts).
- ID fan of bigger size is required due to higher volume of gases handled. Bigger sized shaft and bearings are required (*hence, accurate dynamic balancing of impeller becomes necessary*). The internals of the ID fan need to be protected by suitable rubber/fibreglass lining if corrosive and moist gases are to be handled.
- Diameter of duct and process vessels is also bigger for the same reason.
- The equipment and ducts shall be externally reinforced/strengthened to minimise the chance of internal collapse (implosion) if pressure falls further below atmospheric, e.g. excessive cooling of the condenser.

18.3.2 Plant Operation Above Atmospheric Pressure

- Plant may not need an ID fan if the forced draught (FD) fan is able to generate enough pressure and has required capacity to push through the gases through all units in the plant.
- Size of FD fan can be smaller compared to ID fan
- *The internals of the ID (if provided) fan need to be protected by suitable rubber/fibreglass lining if corrosive gases are being handled.*

18.3.3 Plant Operation at Above and Below Atmospheric Pressures

Some units at inlet (feed) end of the plant operate above atmospheric pressure (i.e. positive pressure) while remaining units operate below atmospheric pressure (negative pressure).

For example, 30 % Oleum is boiled to generate SO_3 vapours at about 125 °C which are sent to the condenser to produce liquid SO_3. Here the pressure can get reduced to below atmospheric pressure (−100 mm WG) on condensation of SO_3. This condenser needs a special dual pressure gauge and pressure relief valve also (to vent of any non-condensibles).

The oleum boiler is also provided with a good pressure relief valve to release excess pressure if the rate of condensation is less due to any reason. The vapours are released through a scrubber irrigated with 98 % sulphuric acid to prevent atmospheric pollution.

18.4 Typical Checklist for Smooth Production Run

Senior plant engineers shall prepare the following checklists so that the planned product mix and delivery schedules committed can be met. These lists will also enable proper execution of any planned maintenance activities. Broadly, the checklist can be grouped as follows:

<u>Daily checks</u> of proper functioning of all important equipment and availability of required inputs at the start of the day and at the end of the day (to ensure smooth run in evening and in night shifts till the next day at least).

<u>Weekly Checklist</u> for smooth plant operation in next few days/next week.

<u>Checklists</u> for further planning and major activities for the future (say next month or till next planned shutdown of the plant). This can include modification of plant units, major repair or replacement of some machineries (which may be possible only during a longer stoppage).

Some considerations for the preparation of these checklists could be discussed.

18.4.1 Coordination with Marketing (Sales) and Logistics Department

To confirm that sufficient quantities of finished product mix and suitable vehicles are available as per delivery schedule programme committed by marketing department to the clients. *This will avoid sudden changes in plant operating conditions if the required products or special vehicles are not available and the commitments made to valued/prestigious clients must be met anyhow.*

- when to position the vehicles/tankers at the loading point in the premises;
- when to load them with filled-up bags/carboys/material.

Plant engineers shall inform if it is not possible to produce the required product as per specifications or in the required amount if there is some problem in the plant. *The delivery schedules shall be then changed rather than trying to meet the delivery commitment by overloading the plant to somehow produce the material which may create safety issues.*

18.4.2 Availability of Items for Process and Maintenance

A certain minimum stock level of items for emergency use shall always be maintained in the plant storage yard/storeroom (near the control room) in consultation with senior chemical, mechanical, and electrical engineers. The quantities shall be sufficient to keep the plant activities running till next supplies are procured and brought from the central main stores.

Production manager may delegate authority to the purchase manager and to the store manager to always ensure the availability of all these items.

An emergency fund shall be available with plant engineer for emergency purchase of spares.

Typical Lists

These lists are only indicative. They shall be revised as per requirement of individual process plants and the machineries installed, and more items may be added.

18.4.3 Daily Checks of All Important Equipment

- Fire fighting arrangements, such as CO_2, and DCP cylinders, are to be in working order. Hose pipes and reels shall be in place.
- Warning alarms, safety devices, and interconnections are in working order. No such device has been bypassed to keep the plant working.
- Confirmation that safety equipment for personal use, eyewash fountains, first-aid boxes, breathing apparatus, gas detectors for dangerous gases (toxic, inflammable), and gas masks are in working order.
- Emergency medical care and ambulance must be in state of readiness always.
- All important units and machineries for feeding raw materials to the process plant are working properly.
- Working of all utilities such as air-drying plant, thermic fluid heating systems, packaged steam boilers, diesel generator sets for emergency power supply,

refrigeration plants, and cooling towers (if volatile liquids are being manufactured or being handled in the premises)
- Packing containers such as bags/carboys in sufficient numbers
- All standby equipment, such as pumps, agitators, to be immediately available
- Availability of space in storage tanks for products being manufactured
- Any equipment/machinery which needs urgent repairs is not left unattended (or allowed to run as such till next day)
- Any repair job is not left incomplete. This may cause accidents in night shifts.
- Instruments and process controllers (control valves) shall be confirmed to be working correctly. Settings shall be as per directive from senior plant engineers/authorised persons.

Sufficient stock of following items shall be always available:

- Raw material and additives.
- Raw water and treated water (for process, boiler feed, cooling tower make-up).
- Catalyst for replacing at least 15–20 % of the quantity required for full replacement.
- Chemicals for process use and for ETP, quality control analysis
- Fuel and accessories for DG sets and furnaces
- Alkali for tail gas scrubbers and effluent treatment plant (*ETP*) (to take care of any sudden load due to process upset), ***and gas scrubbing systems must be in working order***; ***and not bypassed***.
- Provision of covering hoods (canopies on equipment generating dust) and exhaust fans with a dust filter and tall chimney for dust control in working areas.
- All necessary accessories such as spray nozzles, demister pads, candle demisters for air pollution control units.

18.4.4 Mechanical, Electrical, and Instrumentation Spares

- Motors, starters, fuses, relays, ammeters, lighting equipment, and cables and wires for important machineries shall be available in ready to use condition
- Bearings, lubricants, coupling halves, sets of V-belts, flat belts, nut/bolts, welding electrodes and complete sets of welding equipment, gas cutting sets with oxygen and acetylene cylinders along with spare pressure regulators, tools for pipe fitter, high- and low-tension electrical technician, welder, instrument mechanics, and machine fitter trades,
- Gaskets of various materials (rubber, polytetraflouroethylene (PTFE), compressed asbestos for steam lines—if permitted by local regulations)
- Critical instrumentation spares and calibration facilities, standard solutions, connecting cables, sensors, probes, thermocouples, and pressure gauges
- Thermal insulation pads (mineral and glass wool)
- Refractory bricks and castable as well as cements

- Spares such as fuses, starters, (electric motors for critical drive), lighting for the plant
- Spares for mechanical repairs—gaskets, plugs, pipe lengths, gauge glasses, sight glasses, nuts and bolts, and **valves** for various service (water, oil, acid, steam, alkali). These shall be of different sizes and with flanged as well as screwed ends.
- Steam traps of various sizes
- Rubber, polyethylene pipes, and hose-tightening clips.

18.4.5 Preventing Interruptions Due to Adverse Situations

- **No water supply/fuel supply for 3–4 days: sufficient quantities shall always be stored to last for at least one week.**
  Schemes shall be already designed for rainwater harvesting and reuse of waste waters.
- **Product dispatch is not possible either due to transport breakdown OR due to non-availability of suitable tankers (with rubber lining or in stainless steel construction) and storage tanks have got filled up.**
  Alternative arrangement made to dispatch by carboys or by direct pumping to nearby customers or the production rates may be curtailed if nothing is possible. Any pending maintenance job may be taken up or annual shutdown may be advanced if possible.
- **Raw material is not available due to any reason.**
  Any pending major maintenance job may be taken up or annual shutdown may be advanced if possible.
- **Heavy rain or snow fall occurs sometimes at the site.**
  Covers for all motors, starters, and light fittings shall be provided in advance to protect from heavy rains. Plant units may be operated at slightly higher temperatures. Steam jacketing may be provided to pipes to prevent solidifications in pipes. If there are chances of an accident due to ingress of steam in the pipes, then steam tracing lines shall be installed.

18.5 Some Quick Fix Solutions for Plant Operation

Leaks from pipelines, process vessels, and small storage tanks are a common phenomenon in many chemical plants. It is not always possible to change the leaking pipe or the vessel immediately especially during night shifts due to practical difficulties such as insufficient manpower and less visibility. In such cases, it is imperative to stop the leak *at least temporarily* till proper repairs or replacement can be done. It may also become necessary to find some immediate practical solutions to operational or maintenance problems though they may not be workable on a long-term basis. These immediate solutions are to be treated *like a first aid* only.

18.5.1 Items to Be Readily Available

These could be kept in control room or in plant stores and shall be always available to the plant engineer in such situations.

- Quick setting chemical putties (chemical-resistant cements mixed with inert material)
- Castable (high alumina) screened refractory cements.
- Tapered plugs of various sizes made from wood, metals (lead), rubber, refractory, insulating, or acid/alkali-resistant brick—for plugging small leaks.
- Pairs of C clamp (3- to 6-mm-thick steel plates in the shape of the letter C) with bolt holes. These can be made in different sizes (to suit pipelines with different diameters) and width. Paste of acid/alkali-resistant material/cement can be applied inside and then clamped tightly on the leaking spot.
- Temporary sling or chain arrangements for hanging loads
- Temporary supports for vessels which are getting tilted (due to some damage to existing support as a result of corrosion)
- Arrangement for cooling with external spray of cold water with a collection tray below for reusing the water
- Steel-braided hose pipes with stainless steel nozzles inserts **OR** polytetraflouroethylene (PTFE)-lined pipes for handling very corrosive fluids
- Buckets filled with coal ash, lime powder, or inert material to prevent spreading of spilled corrosive liquids.
- Electrical heating tape to be wound on a pipeline for warming it (this should not be used for inflammable fluids inside).
- Vacuum cleaner (aspirator) for sucking out gases from a vessel
- Hairdryer (used by many women) for quick drying of electrical contacts if some moisture is seen and it is to be removed quickly before further repairs.
- Small portable tripod stands with provision of chain pulley blocks of 0.5, 1.0, and 2.0 MT lifting capacity can be used for providing temporary supports to sagging pipelines/to process vessels (where one or more support has got broken)
- Raw material may be filled in small bags and lifted up by means of chain pulley blocks to the charging manhole of the reactor if the hoist has broken down.
- Inlet nozzle for feed water of a process unit has got choked and it is not possible to open and clean it without stopping the unit. In such a situation, supply feed water through the vent line for preventing the water level going down to a dangerous level.
- Add ice slabs (*if available from nearby units*) for chilled/cooling water system if it is required to immediately control the temperature in a vessel handling inflammable/toxic vapours and there is a breakdown in the existing refrigeration system. This is a costly method—but it may be able to prevent an accident in the plant.
- In case of the regular boiler not able to supply steam due to some reason, a small capacity packaged quick-start-type (sometimes called *baby boiler*) boiler shall be available. This is for supply of a small amount of steam at low pressure only.

18.5.2 A Word of Caution

While using these quick fix solutions, the plant engineers must consider the safety issues involved and shall never exceed the upper safety limits on operating conditions for the particular equipment (speed of the agitator or operating pressures, temperature beyond specified by the manufacturer or instructed by statutory authorities).

- It is to be clearly understood that the clamps put on a leaking pipe are as a temporary measure only and the leaked pipe must be replaced immediately or at the earliest. There can be a possibility of a repeat leak from the same spot and hence the area in the proximity of such clamped pipes shall be cordoned off and warning signboards must be displayed. The clamps shall be kept covered so that no spray of the liquid can come out if it leaks again.

18.6 Equipment Modification for Adding Cooling/Heating Capacity

It may become necessary to increase the production rate for meeting some urgent demand in the market. It will be quite costly and time consuming if the cooling (or heating) capacity is to be increased by installing a new reactor or major modification of the jacket is to be carried out. In such situations, the following may be considered.

For additional cooling,

Run both working and the standby cooling water pumps together if a higher flow rate is required and higher capacity pump will not be available soon. This will be useful for circulating greater quantity of cooling liquid in a system immediately.

Use chilled water (if available) instead of cooling tower water.

Explore if agitator can be run at higher speed, blades can be changed or incoming reactants can be precooled.

For additional heating

Use higher pressure/temperature steam or a heat transfer oil (thermic fluid) at higher temperature for heating provided the jacket and the reactor shell are able to withstand these conditions. Check suitability of high temperature for the reaction mass inside before hand.

Addition of external heat transfer surface

(*When heat transfer area of jacket is not sufficient*)
Provision of an external recirculation arrangement through a heat exchanger (with a small pump) will be found useful in case it is not possible to change the reactor. This will not disturb the existing reactor operation (Fig. 18.1).

This will be found useful to augment the cooling/heating capacity of the reactor if the production rate is to be increased with minimum additional cost—since the same reactor may be used.

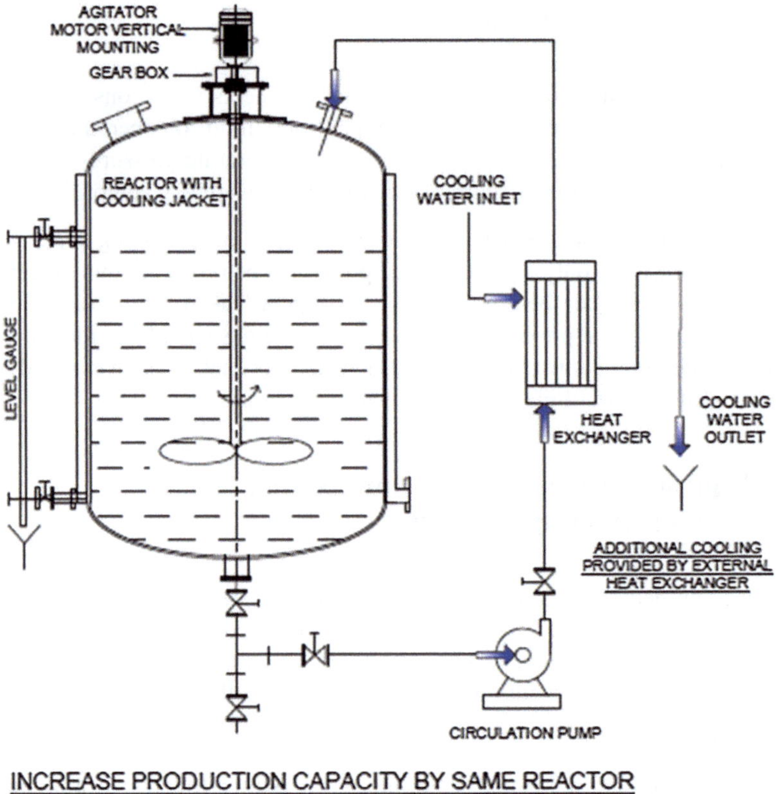

Fig. 18.1 Addition of external heat transfer surface to increase production capacity by the same reactor

This may be found useful even in the case of damage to the agitator (if it is going to take a long time to repair the agitator) in the reactor and at least some production should be continued.

18.7 Selecting the Process and Technology

It may be possible to produce a chemical by different processes or technologies. The one which is most efficient and safe to operate shall be chosen. The selected process and the technology should be latest and suit the local geographical conditions, easily available raw materials (preferably from local sources). It should be safe and easy to operate.

- Consumption of raw materials, power, steam, water shall be lowest or comparable with the best in industry. Specific consumptions of all inputs per unit of output shall be very economical over all operating capacity ranges.

- Dependence on imported items (or *brought from distant sources*) shall be minimum.
- Material of construction of process units shall not contaminate the products and should ensure long life of units.
- Equipment shall be easy to clean, maintain, and repair.
- Easy to operate heating and cooling methods,
- Automation of process (there shall be less reliance on human elements)
- Reliable instrumentation and automation shall be incorporated, with arrangement for manual override if required.
- Water-cooled units shall not be used in a plant where water is not sufficiently available at the plant location (or the quality of available water is bad)
- Avoid selection of very complicated technology which is difficult to understand and there may be chances of mishaps in the case of operational mistakes.
- It shall be possible to operate the plant efficiently and safely at varying production rates as per market demand (if higher demand for product is to be met urgently). The overload operation shall be possible up to 120–130 % of rated capacity without occurence of dangerous situation.
- It shall be possible to diversify to other products by small changes in the process units or addition of balancing units.
- The products shall be easily marketable, and the quality should meet the specifications (domestic and international) at the minimum and maximum production rates. Generation of off-spec products shall be minimum.
- The break even point (BEP) shall be at as low capacity as possible.
- Sufficient warning alarms, safety devices, and interconnections must be incorporated and vendor shall demonstrate satisfactory working of all such systems prior to handing over the plant to purchaser.

These arrangements shall never be bypassed while running the plant.

- The plant should not cause any environmental pollution. Waste generation shall be minimum during start, normal running, and also during annual shutdown of the plant. There should be in-built effluent treatment facilities to treat all effluents.
- Critical pieces of plant and machinery shall have standby units installed at the right locations such that they can be immediately taken into use.
- Contract for purchase of plant and machinery shall include conditions for technical guidance for at least one year.
- Visit a satisfactorily running unit based on similar process and technology before accepting a plant based on the process and technology being offered for selection.
- The process units shall be designed and fabricated as per codes and standards to be statutorily complied with in the country *where the project is being set up.*

All statutory requirements and instructions shall be complied with by the owner.

Chapter 19
Technocommercial Matters

19.1 Revival of Plants

Marketing organisations generally have good information about the demand for various products, their specifications, and trend in prices. They may become desirous of manufacturing these products themselves if they sense a good business opportunity. Manufacturers of chemicals also have their own marketing departments and may think of increasing production capacity of their own plants or diversify to new products since they already have good information about the demand.

Managements may have to choose from the options such as expanding the production capacity of the existing plant, add new process units in the same premises, and set up a new unit elsewhere or revive an old plant which is *not running profitably or has been idle for long time.*
(*the expansion of production capacity of the existing plant or diversification to new products by same planthas been addressed elsewhere in this book*).

19.1.1 Setting Up a New Industry

Setting up a new industry from grassroots level needs many activities such as acquiring land, obtaining connections for power, arranging water supply, finalising process and equipment design and procuring them, construction of storehouse and civil foundations, and erection and commissioning of process plant.

K.R. Golwalkar, *Production Management of Chemical Industries*,
DOI 10.1007/978-3-319-28253-4_19

19.1.2 Revival of a Plant

Revival of a plant may or may not need all these. It may appear attractive to invest for revival of an old plant especially if the price quoted for it is quite low. However, before investing funds in such venture, one should find out the reasons why it was idle or stopped.

Some probable reasons could have been:

- Raw materials (as per specifications) were not available.
- Sufficient utilities could not be arranged regularly (water, fuel, power).
- Plant operations were not stabilised soon, and rated production capacity was not reached for a very long time (even after many modifications or replacements were carried out during trial runs and commissioning itself).
- Technology had become obsolete by the time the plant was ready for start up. Hence, this plant could not compete with more modern plants which had entered market with a better/cheaper product in the meanwhile. Moreover, inefficient operating equipments were wasting raw materials and power. *Discard the present technology if it has become obsolete and better machines are now available which can run with less breakdowns and with less manpower.*
- Breakdowns were frequent (reasons need to be investigated before takeover).
- Poor maintenance—critical spares were not arranged in time.
- Production rate was low (hence high cost of production) and/or product quality was poor, and hence, profit margin was very less due to low sales realisation.
- Many mishaps/incidences of environmental pollution took place, and permission to run the plant any further was refused by statutory authorities *in present condition of the plant.* Many of the existing units may have to be disposed off in this case.
- There were labour problems.
- Major damage had occurred due to natural calamities (storms, earthquakes, floods).
- Batch production was carried out, while competitors had continuously running plants.

The technical and economic problems could have been due to incorrect specifications, long delivery schedules, non-availability of some critical components or spares in time, mistakes made in erection or commissioning, faulty equipments having frequent breakdowns created difficulties during operation and maintenance, and sufficiently trained workforce not appointed in time.

- Situation could have got aggravated further in case the production run was not stabilised soon or the finished product was not saleable since it did not meet the required specifications.
- Typically, in such cases, it becomes necessary to store the off-spec product till the time the product meets the specifications given by customers. The lower-grade product will have to be sold at a much lesser price (sometimes even at a loss) or reworked upon for improving the quality (if possible by recycling or

other means) or disposed off in some other ways acceptable to statutory authorities. *This delays generation of revenue and lowers the rate of earnings from the industry.*

- Meanwhile, consumption of raw materials and other inputs goes on specially in case of a continuous process plant. *However, the essential purchase of fresh supplies of raw materials and other inputs need more fresh loans.*
- A severe shortage of working capital during this period can also force closure of the plant. Further, the instalments of loan and interest cannot be paid on time to financial institutions.

19.1.3 Site Visit and Inspection

It is necessary that a visit is undertaken to the site by experienced senior engineers *to assess first hand*:

- The present condition of the process units and machinery and civil structures,
- To check the consumptions of raw materials and utilities in the past,
- To obtain samples of raw materials which were being used in the past to check its purity and level of contaminants,
- To examine the history of any mishaps that had taken place,
- To study the adverse comments made by statutory authorities,
- To examine general state of installed machines,
- Condition of warehouse for raw materials and finished products—this is important because storing in the open can spoil the materials if a properly built storage space is not available,
- Whether sufficient water of the proper quality was not available and supply can be arranged now?
- Records of the operating conditions, product quality, production rate, and consumptions of raw materials and utilities should be asked for if available. These shall also be considered for the modifications/upgradations/procurement of balancing equipments and shall be studied by consultants along with own engineers (who should also give their opinions and suggestions) and any other relevant observation.

19.1.3.1 Other Matters to Be Observed/Checked

Unpaid loan or interest instalments to financial investor institutes and stakeholders shall also be carefully studied because fresh loans may be denied till some stringent conditions are fulfilled.

Any unpaid bills, government taxes, and dues to workforce shall also be looked into.

19.1.3.2 Detailed Inspection of Old Equipments

The detailed inspection of all units shall be done with due reference to the above information. A typical list of activities and points for diligently looking into such inspection can be:

- Mothballed Plant: whether proper precautions were taken prior to shutting down in the past. e.g., whether allresidual chemicals from reactors, condenser, heat exchangers were drained out completely; the units flushed byappropriate means (alkaline solution, nitrogen gas etc) and then preserved carefully.
- Opening and cleaning of internals of process units such as absorption towers, heat exchangers, reactors, and condensers—*if necessary now* (*before revival*),
- Servicing of machineries, which are in working condition—(pumps, blowers, etc.),
- Repair of equipments and machines which can become usable thereafter—bucket elevators, agitators in reactors, and hoists,
- Upgrading some units such as plate heat exchangers by adding more plates and converter by adding more catalyst,
- Replacement of damaged units which cannot be repaired—damaged valves, instruments, demisters, corroded pipelines, and ducts.

19.1.3.3 History Cards of Equipments and Machineries in Idle Plants

History of operating conditions, problems that took place, and maintenance work done shall be studied for deciding on procurement of new items/major repairs/modifications.

The history may not be available for an old plant lying idle.

The following shall be analysed before deciding to buy new equipments in place of old ones:

- Serial number, year of manufacture/procurement, and date of commissioning,
- Whether standard operating conditions as given by original manufacturer were followed or equipment was overloaded?
- Whether standard repair procedures were followed? Whether the spares were always procured from authorised stockists/original manufacturer or reconditioned spares were used?
- Where was the equipment actually installed (as per the design or in a dusty, hot, outdoor location)?
- Actual working conditions (dusty surroundings, outdoor installation) and deviations from design duty conditions,
- Original general assembly drawing and battery limits specified,
- List of spares and their drawings and code numbers,
- Duration and reasons for the unit being idle.

19.1.4 Further Planning

- Consultants having necessary expertise and experience can be engaged for working out scheme for revival. Also organise a special team with own technical, marketing, and finance experts for coordination with the external consultants. Senior engineers and experienced workers (if available) shall also be included in this team for consultation.

19.1.4.1 Basic and Detailed Engineering Documents

Basic and detailed engineering documents should be prepared jointly for the revival, expansion, diversification, and/or modernisation of the old units. Consultants shall provide the know-how, list of necessary equipments and machines, their specifications, and assistance for procurement. Balance facilities required may be fabricated in-house by own engineers, fabricators, and technicians if possible. The company can modify existing units in small incremental steps with assistance from the consultants and technical teams.

They shall make small changes to or provide additional features to existing machinery to obtain required performance with small incremental investment at a time. This can minimise additional capital required and can make maximum use of existing units. In certain cases, this approach will not work, and it will be necessary to make major changes even to start the plant (e.g. changing the complete reactor itself).

19.1.5 Requirement of Funds

An estimate of funds required and schedule for cash flow of funds shall also be prepared.

(i) for revival while manufacturing the same products initially,
(ii) attaining rated capacity on stabilisation,
(iii) expansion of production capacity, and
(iv) diversification to more products if possible.

- The estimate of the funds required for the proposed corrections or repair of equipment/replacing some machines shall be reasonably correct. Provision of about 8–10 % additional amount as contingency funds should also be made.
  It shall thus need much less capital to repair and start the plant instead of buying a completely new plant. This can save on considerable interest on the large investment (or loan from financial institutes) which is required for a new plant.

This team shall obtain **performance guarantee** from consultants that the plant will be able to run safely, at the required production capacity, quality, and efficiency, with guaranteed lower consumptions of raw materials and utilities, without causing environmental pollution.

The production rate after revival may be less initially (but shall not be less than the original rated capacity as far as possible). It shall be possible on stabilisation to increase it further so that cost of production per unit can be brought down.

It should thus become possible to win back the old customers and get new ones also.

19.1.6 Revival with Same Products: Process and Equipment Modifications Only

Study presents process, equipment design, and operations carefully from the following angles:

- Was the degree of conversion sufficient? If not, some of the reactants may be getting lost since they may not be getting converted to required products (due to loss of activity of catalyst, less residence time in reactors, poor heat transfer in process units, poor mixing of reactants, and side reactions taking place due to poor operating methods).
- Did the process operate with very high pressures/high temperatures/special materials of construction/had complicated operating steps? ***Simplify the process and make it safer to operate***.
- Was it a batch-wise or a continuously operated plant? (there could be variations in product quality in batches; cleaning of reactors or other units may be required for every batch—this can reduce production capacity and cause environment pollution due to washing of reactors as well as waste of reactants).
  Longer production runs are preferable instead of short production runs which need frequent stoppages of plant. These waste more fuel, power, and manpower every time the plant is started and stopped—moreover, the rate of production is less during starting and stopping at least for a few hours. Hence, try to minimise idle times between batches or convert present batch operations to continuous operations.

Change of Process or Technology: Case study

- Manufacture of refractory bricks by baking in small batches was not economical. Lot of fuel was wasted, and time was required to heat and then cool the batches. Hence, the baking operations were then carried out in a long tunnel furnace with heat recovery (air preheating using heat of hot bricks from exit end).

- Manufacture of sulphuric acid by the modified 3 + 2 DCDA process (having three catalyst passes before interpass absorption tower and two passes after it) with caesium-promoted catalyst resulted in better conversion of SO_2 to SO_3 and negligible escape of SO_2 from exit gases. An alkali scrubber for tail gases was provided for taking care of this.

The cost of production could be considerably reduced in both cases.

19.1.7 Raw Materials

Does the plant need a raw material with very high degree of purity or of certain specifications which are not easily available at reasonable cost? Can some other raw materials be used?

Are they to be procured from some distant source?

Check landed cost at site for other raw materials plus transport cost.

Compare cost of material from different places with the transport cost. Check reliability of supply if procurement is to be done from long distance.

19.1.8 Cost-Cutting Without Compromising on Safety

- Check every item which contributes to the cost of production.
- Reduce/stop all wasteful activities, reduce inventories of raw materials and spares, develop cheaper spares in-house, better conversion of inputs to saleable products, reduce polluting streams/recycle or sell the wastes generated if possible.
- Replace old inactive catalyst by fresh catalyst as per advice from manufacturer for loading in the existing plant design.
- Thoroughly revamp the effluent treatment facilities and get them approved from statutory authorities.
- Implement process control programme at every step of production (analysis of raw material, reactor operation, calibration of instruments, replacement of catalyst, filtration of raw material/product streams, purification steps for final product, etc.).
- The selected process technology should be latest and shall be able to stand competition from other manufacturers.
- It shall be possible to operate the plant efficiently and safely below rated capacity (when the demand for products is less, raw materials or utilities are not available) or at 120–130 % of rated capacity when higher demand for product is to be met urgently.

19.1.9 Revival with Expansion of Capacity

Production capacity shall be increased on revival—only after stabilising operations at the original rated capacity so that proper functioning of all units has been established.

This can be considered if there is strong indication from marketing department that the demand will definitely increase soon or in near future.

- This will put additional *operating load* on most of the process units (reactors, condensers, absorption tower, filtration systems).
- Other existing facilities such as water treatment, main transformer and electrical cables, bus bars, motor control centre, cooling tower and chilling plants, fuel storages and handling systems, fuel burners, boilers, and material handling systems such as screw conveyors and bucket elevators will also have to operate at increased load.
- Specially, the capacity of air and water pollution treatment equipments must be checked to handle the additional load of effluents without which permission will be denied by the statutory authorities or the plant expansion. Equalisation tank capacity will have to be checked/increased.
- Upgrading/reinforcements of all civil structures and other structural supports may also be required due to additional weights of more equipment.

19.1.9.1 Certain Equipments Shall Need Upgradation

Certain facilities and machineries like Storage Tanks for raw materials and finished products, overhead cranes, refrigeration plant, bucket elevators, screw conveyors, equalisation tanks of Effluent Treatment plant (ETP), fuel storages, cooling water pumps, Motor Control Centre (MCC), certain electrical cables may need upgrading.

Safety of personnel and plant units must be the most important consideration while expanding production capacity and diversification to new products. There shall be need for additional manpower for operation of plant and auxiliaries, maintenance, handling of raw material and finished products, filling up tankers, inventory management, etc. Only qualified and experienced new persons shall be engaged.

19.1.9.2 Providing New Equipments for Revival/Increasing Production Capacity

It may become necessary to procure new equipments even after carrying out certain corrective actions (*some typical ones are given below*) and upgrading existing units to the maximum extent possible. Typical areas could be effluent treatment plant, electrical cables, MCC and motors, water treatment, refrigeration plants, bigger pumps and compressors, heating facilities, additional storages, material handling, etc.

- Remove all bottlenecks—servicing, upgrading, and repairing of old equipments.
- Ensure the safe running of all equipment at present and at increased load.
- Check whether the process can be operated at less severe conditions of temperature, pressure, acidity, etc., to reduce breakdowns or use better materials of constructions for critical parts when the capacity is to be increased.
- Set up grievance cell to attend complaints from labour.
- Carry out steps to minimise energy consumption and maximise energy recovery because the energy losses could increase if all equipments are not optimally increased in capacity—(since some old equipments may be only serviced and retained as such).
- Overhaul process units thoroughly by adding better catalyst, clean heat exchangers, filters, absorption towers, repair refractory lining, and ETP.
- Reduce idle time between batches and convert batch operation to continuous operation.
- Implement modern maintenance practices of condition monitoring and preventive maintenance and strengthen equipments prone to breakdowns.

19.1.9.3 Time Required

The time required to revive, expand and/or diversify depends on activities like dismantling of old units; cleaning, repairing, upgrading some units, buying new units, erecting and commissioning new equipments as well as the time required for training the workforce to operate and maintain the various equipments. Try to retain as many old units as possible to minimise cost of dismantling.

It shall be noted that no revenues will be generated till the plant starts supplying finished products to customers—however, this time period can be less as compared to a completely new unit being set up.

19.1.9.4 Diversification to More Value-Added Products

This can be considered in the following situations:

- There is definite information about demand for the products at present and will increase in future.
- These products will add to the profits of the company.
- Existing infrastructure is either sufficient or will need only some balancing additions.
- Many of the surplus capacity of existing units can be used (refrigeration plant, filtration, water treatment, material handling, etc.).
- Some of these products can be part of vertical integration in the organisation, e.g. bamboo pulp → rayon fibre → blended cloth → dress materials,

Aromatic plants → steam distillation → essential oils → medicinal formulations,
Sulphuric acid → 25 % Oleum → Liq SO_3 and 65 % Oleum; Alum; Superphosphate.

19.2 Technical and Commercial Audits

The technical audits shall be done by an objective inspection of the plant and machinery as well as the infrastructure *in order to confirm* that the following activities and jobs have been executed properly.

19.2.1 Planning, Design, and Fabrication Activities

- Market study and analysis of demand for the products:
- It is required for the planning of product mix (which is necessary for design of process units for the present and to ascertain feasibility of the plant in the future). *This should have been done with due diligence.*
- Technical specification preparation has been done by competent professional experts who have sufficient experience of designing, operating, and maintaining such plants with similar product mix and who have the necessary authorisation.
- Quality Assurance Plan obtained will be able to meet the desired standards for: *Material of construction, methods of fabrication, stress relieving, various tests of the equipments at every stage of manufacture as required for safe and satisfactory performance.*
- Experience of fabricator/supplier (for the project) is professionally adequate for meeting all technical requirements.
- The technical equipment *being procured* will be designed and fabricated as per applicable (international) codes and standards and will be suitable for the operating conditions.
  These codes should be acceptable by statutory authorities in the particular country where the equipments will be installed.

19.2.2 Procurement and Purchase Activities—Legal and Commercial

- All legal issues regarding procurement (especially for procurement from abroad) are complied with. These shall include conditions for transfer of know-how and

supply of components made from special materials of construction (which may need licence from vendor's country).
- All import duties and government taxes shall be paid, and clearances from customs shall be obtained. There shall be no delay for procurement of any of the items due to legal formalities required for export/import.
- Proposals for commercial matters and offers from all vendors shall be objectively evaluated by independent professionals before making final choice.
- Financial matters shall also be evaluated likewise.

19.2.3 Regarding Implementation of the Project

Completion of various project activities in time/as per scheduled dates is a very important matter. Typical activities relate to obtaining all statutory clearances, acquiring land, arranging water and power supply connections, completion of civil foundations and other infrastructures, procuring necessary plant items and machinery, and their erection and commissioning, followed by stabilising the production runs.

The performance guarantee test (PGT) runs must include reaching rated capacity of the plant with acceptable quality of products without exceeding consumption norms (and any other conditions agreed between the parties) on or before agreed dates.

It is advisable that the PGT runs for a batch-operated plant shall be on the basis of five batches of the product, while the PGT runs for a continuously operated plant shall be based on 72–168 h (three to seven days) of continuous run before signing the acceptance test report by both parties (all other conditions as per contract shall also be met). Balance payment may be released only after satisfactory PGT runs.

The methods for measurement of production rates, consumptions of raw materials and utilities, and compliance with effluent control norms should be agreed between vendor/consultant and plant management before conducting the PGT and shall be recorded in the agreements. Penalties may be prescribed for not meeting the terms of PGT.

There shall be a clause in the contract between vendors and purchaser for making up the liquidity damage in case of delay in project completion.

19.2.4 Impact of the Project on the Organisation

The expected benefits from the project are generally defined as:

- Safer operation and better pollution control.
- Increased efficiency leading to lesser operation cost and better product quality.

- More acceptability in market—repeat orders from old customers and getting new customers due to building up of reputation.
- Increase in income for the organisation with a benefit–cost ratio acceptable to the management.

19.2.5 Appointment of the Technical Auditor

The auditor should be an expert with knowledge of both the technology and the market and needs to be completely independent. He should objectively assess the situation by asking searching questions to the plant personnel and also make independent observations.

Since the auditor is allowed access to almost all internal information, he must maintain the strictest confidentiality regarding all proprietary information.

Plant managements shall also carry out the following additional audits to ensure proper working of the production activities.

19.2.6 Safety Audits

To confirm that proper process controls are in place to minimise the possible risks to personnel, plant equipments, and surroundings (environment).

- All necessary safety devices (safety valves and rupture discs), automatic activation of safety systems if any dangerous condition develops, and safety interlocks are confirmed to be in working order. There shall be no chance of any tampering of the safety systems.
- Safety audits shall also include personal safety items, safety showers, eyewash fountains, dedicated overhead water tanks, and fire fighting systems,
- Electrical earth connections, short-circuit protections, etc.
- Inspections and maintenance of equipments are being carried out regularly.
- All pressure vessels, boilers, high-pressure pipelines, etc., are regularly tested in accordance with statutory rules, and any leak/damage is attended immediately. Subsequently, the item is tested again before taking into use.
- Work permit procedures should be followed. Safety organisation shall be in place. Any mishap shall be investigated thoroughly and the reasons determined. Suggestions for corrective actions must be objectively evaluated from technical angle also.
- HAZOP studies shall be carried out whenever any new equipments / operating procedures are introduced**or** every three to six months, whichever is earlier.
- Training programmes for personnel, *i.e. the plant operations are being performed as per standard operating procedures with revisions, if any.*
- Authority and responsibility for operating activities are assigned properly.

- Information system is in place to confirm that operating activities being performed properly.
- Any other safety arrangement provided by plant engineers/instructed by statutory authorities to be installed.

19.2.7 Performance Audits

An objective analysis shall be done to confirm the following:

- Plant is running at rated production capacity.
- Quality of products is as per specifications/as required by marketing department.
- Consumption of all inputs is as per standard norms with special attention to power, steam, and water.
- Quantity of effluents generated is within control, and they are discharged only after proper treatment.
- No process unit or machinery is being overloaded beyond its rated capacity/more than recommended load by supplier.
- No safety device is bypassed or deactivated for convenience of operation.

19.2.8 Energy Audit

Chemical plants mainly consume thermal and electrical energies. An energy audit of these plants should examine closely not only the consumptions of these but also the conservation and recovery of heat as well as cogeneration of energy in these plants. All these aspects are covered in the chapter on Energy Efficiency (Fig. 19.1).

19.2.8.1 Important Stages of Energy Audit

Important stages of energy audits are as follows:

1. **Collection of data for** (i) all consumption of electricity, steam, coal, furnace oil, gaseous fuels for the plant and diesel for the generators. (ii) all power inputs such as power from grid and self-generation as well as all thermal energy inputs and (iii) all energy recovery sources.
2. Compare with norms/original guarantee figures as given by vendors.
3. Field study of the plant and individual equipment to compare actual working conditions with original specification of the individual equipment (e.g. heating/cooling requirements, mechanical drives provided, their running timings).

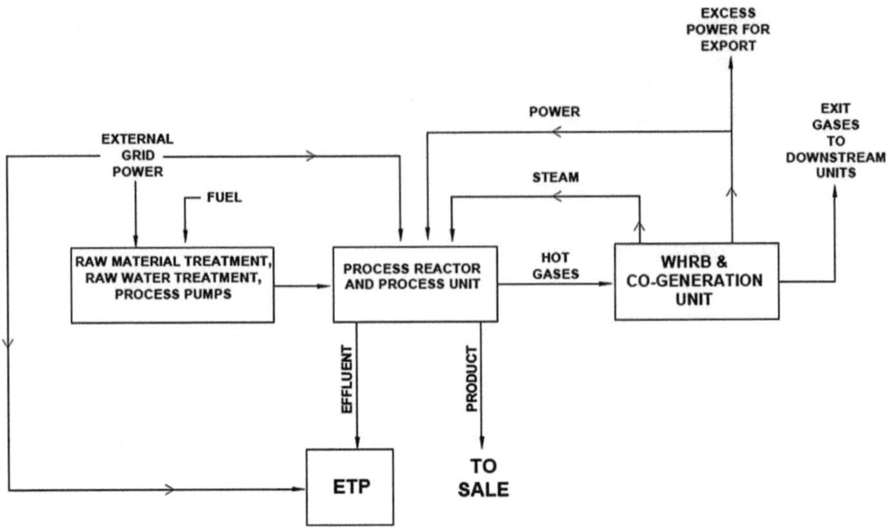

Fig. 19.1 Overall energy scenario for the chemical plant

4. Explore possibilities for improvement by better thermal insulation to reduce heat losses; by variable frequency drives to run the machinery at optimum speed, use of light-emitting diodes instead of large bulbs; by immediately attending to any compressed air leak in the plant and faulty steam traps; modifying the material handling procedures and reducing the running hours of all equipments consuming more energy.
5. Preparing a draft report with analysis of actual conditions, suggestions for improvements, possible benefits, investment, and time required for implementation with payback. This shall be discussed with shop floor engineers to know their views and to find out any genuine difficulties faced by them before presentation to senior management.
6. Discussion with senior management and preparing a time-bound programme for implementation for the matters approved.
7. Following up of the programme and objective assessment of the results obtained.

19.2.8.2 Regular Examination by Thermographic Cameras

This is useful for detecting the heat losses from hot surfaces. The images shown by the camera indicate the presence of hot spots on surfaces of ducts and pipes (indicating inadequate or damaged insulation), furnaces, and process units (indicating inadequate or damaged internal refractory layers). The observations can be made quickly and even from a distance up to 10–12 m from the hot surface which makes it safer also.

These cameras can also detect loose electrical bus bars and loose joints in cables carrying high currents because these can become quite hot. Immediate corrective actions of cleaning and polishing the surfaces and retightening the loose portions shall be taken. These can prevent electrical breakdowns as well as save electrical power due to loose connections. *Incidences of fires can also be reduced as an added advantage.*

- It is also possible to detect a bearing or a machine part getting very hot (impeller rubbing against the body of a compressor) and to take suitable corrective actions in order to prevent mechanical breakdowns.
- Do not generate steam unnecessarily at a high pressure in the boiler and then use a pressure reducing (controlling) station (**PRS**) for bringing down the pressure at point of use. *These installations also have pressure relief valves (PRV) at downstream to protect the downstream equipments in case of malfunctioning of the PRS.* Instead, generate steam at just sufficient pressure and use it as such. This will reduce exit gas temperature from the boiler since the saturation temperature of lower-pressure steam is also lower.

Index

K.R. Golwalkar, *Production Management of Chemical Industries*,
DOI 10.1007/978-3-319-28253-4

O

P

MIX
Papier aus verantwortungsvollen Quellen
Paper from responsible sources
FSC® C105338

If you have any concerns about our products,
you can contact us on
ProductSafety@springernature.com

In case Publisher is established outside the EU,
the EU authorized representative is:
Springer Nature Customer Service Center GmbH
Europaplatz 3, 69115 Heidelberg, Germany

Printed by Libri Plureos GmbH
in Hamburg, Germany